THE HANDBOOK OF CHEMICAL SUBSTITUTES

A Handbook of Substitutes and Alternatives
for Chemicals and Other Commercial Products
Including a Plan for Making a Proper Choice

by

H. BENNETT, F.A.I.C.

Editor-in-Chief—The Chemical Formulary
Director—B.R. Laboratory, Miami Beach, FL 33140

CHEMICAL PUBLISHING CO., INC.
New York, N.Y.

The Handbook of Chemical Substitutes

ISBN: 978-0-8206-0084-0 Paperback

Chemical Publishing Company:
www.chemical-publishing.com
www.chemicalpublishing.net

First Edition:

© **Chemical Publishing Company, Inc.** - New York 1985

Second Impression:

Chemical Publishing Company, Inc. - 2011

Printed in the United States of America

TABLE OF CONTENTS

BOOKS BY H. BENNETT

The Chemical Formulary Vols. I–XXVI
Concise Chemical & Technical Dictionary
New Cosmetic Formulary
Chemical Specialties
Industrial Waxes Vols. I & II
Practical Emulsions Vols. I & II
Trademarks
Encyclopedia of Chemical Trademarks
More For Your Money

Preface

Historically, this book was started about 40 years ago. At that time, the cosmetic, drug and flavor industries were seeking substitutes for glycerin and ethyl alcohol—for economic reasons and to avoid the red tape connected with the buying, storage, use, and selling of alcohol and alcoholic products. Fair substitutes were developed for both of these products, but when the price of glycerin was stabilized at a reasonable figure, these substitutes were almost completely forgotten.

Over a period of years, this writer has developed substitutes or alternatives for numerous products in diverse industrial fields. Thus, a file of such materials has been built up. To this has been added the suggestions of others and references from scientific and technical journals and texts.

This book cannot be regarded as complete or encyclopediac. The subject matter is in a state of flux and is growing and changing continuously. It should be useful to many as a starting point. It should not be expected to give the final answer to a highly specialized need. It is the task of the specialist or expert to glean from it what may be applicable and to interpret, interpolate, or "imagineer" a solution to his specific problem.

Condensation, rather than elaboration, has been the precept in assembling this information, in order to expedite the publication of this book. It is the concentrated essence of many years of experience of many chemists, engineers, and other technical workers.

This writer will greatly appreciate learning of any errors, omissions, or additions that might be made from those who use this book.

H. BENNETT

v

NOTICE

Chemical specialties or proprietary products mentioned in this book by their brand, trade names or trademarks, will be indicated by the use of quotation marks. This book is sold and may be used only on the basis that no responsibility is assumed by the author or publisher in connection with these products whose names are commonly used by chemists, engineers and others.

Introduction

Whether they be called substitutes, surrogates, replacements, or alternatives, such materials have been used from time immemorial. Sometimes they have been used to reduce costs; sometimes to replace unobtainable materials; and sometimes to produce better or different properties.

Selecting the proper substitute is no easy task. Since no material has all the same properties as the material which it will replace, it cannot be expected that the replacement will yield a finished product possessing exactly the same characteristics as the original. A replacement, therefore, that will produce a finished product which will perform almost the same function as the original, without too great a difference, is ordinarily considered satisfactory. For example, glycerin, in an antifreeze, has been satisfactorily replaced by ethylene glycol even though the two products differ in certain chemical and physical properties.

A substitute material, excluding price and availability, must be considered from many angles before it can qualify as a good substitute. Since it cannot have *all* the same physical and chemical properties as the original material, a compromise must be made. Thus, corn syrup may be suitable as a glycerin replacement in a suspending medium, where its viscosity is primarily desired, as in certain toothpastes. It is not of importance that corn syrup does not lower the freezing point of water or that it is not as hygroscopic as glycerin. Where, however, the last two factors or others are important, the use of corn syrup in place of glycerin, is not advisable.

Even when a suitable substitute is found, it may be necessary to modify the original formula by using a smaller or larger amount of the substitute and often, to add one or more other ingredients to balance it. Thus, because corn syrup is more viscous and less hygroscopic than glycerin, it may be necessary to reduce its viscosity by the addition of water and increase its hygroscopicity by means of a compatible hygroscopic salt. Introduction of these two additional ingredients may require considerable testing and aging to avoid subsequent undesirable effects.

Because of the uncertainty of the continued availability of any substitute material, it is advisable to try out a number of materials on each problem, so as to have a substitute ready for the substitute used. It means additional work, but is a worthwhile insurance for continuance in business.

1

Sometimes it may be desirable to change the composition of a formulation radically or entirely because a suitable substitution cannot be made. For example, flavoring extracts depend on the use of pure alcohol as the solvent for the flavoring ingredients. Since there is no good substitute for alcohol (in food products) available, a formuation without alcohol is indicated (e.g., lemon oil) made with an edible gum (gum tragacanth) and water. Of course, the finished product does not look like the original lemon extract, but it can replace it in most of its uses.

Price should not be too great a deterrent in selecting a substitute. Sometimes a substitute will alter a product so as to make it more useful, desirable and salable. An example of this is the use of monoglycollin in place of glycerin. Although the former is more than twice as expensive as the latter, its much greater solvency for certain dyes makes it far more economical to use than the cheaper material which it will replace. In electrolytic condenser manufacture, mannitol, at about three times the cost of glycerin, is replacing the latter because it gives a much more desirable product.

In getting outside assistance in finding a substitute, it is important to disclose a problem in its entirety. Reputable manufacturers and consultants hold all communications in strict confidence. Therefore, give them the complete formulation, method of manufacture, packaging, and a sample of the finished product. Also inform them how and where the finished product is to be used. Only with such complete information can an intelligent recommendation be made.

Substitute Requirements

Every chemical is unique in its chemical and physical properties. Therefore no chemical can replace another equally in all its characteristics. The following list which should be scrutinized gives most of the factors which must be considered in searching for a suitable substitute or alternative. Only those properties which are absolutely essential should be demanded; otherwise the search for a substitute will be greatly hampered, if not made futile.

Adhesion
Adsorption and Absorption
Availability
Bacterial Content
Boiling Point
Caking or Agglomeration
Carcinogenicity
Color
Compatability (Interaction)
Corrosiveness
Cryoscopic Properties
Density and Specific Gravity
Dispersing Properties
Drying Qualities
Edibility
Effect of Aging
Effect of Microwaves
Effect of Soundwaves
Effect of Ultrasonics
Effect on Animals and Plants
Effect on Bodily Functions (Other than Poisoning)
Effect on Skin, Hair, or Fingernails
Effects of Pressure
Elasticity
Electrical Properties
Emulsifiability

Explosiveness
Feel or "Hand"
Flexibility
Form
Freezing Point (Melting Point)
Gelling or Thixotropic Tendencies
Grade or Purity
Handling
Hardness
Heating Power
Homogeneity
Hygroscopicity or Efflorescence
Impurities
Inflammability
Interaction with Other Materials and Containers
Legal Restrictions
Length
Lighting Power (Candle Power)
Odor
Optical Properties
Patent Infringement
Plasticity and Ductility
Plasticizing or Flexibilizing Properties
pH
Polymerization

Radioactivity
Resistance to Oxidation
Resistance to Shock
Resistance to Ultraviolet
Resistance to Vibration
Slipperiness or Friction
Solubility
Solvency
Sound Conductivity
Stability
Static Properties

Stickiness
Sublimation
Surface Tension
Taste
Tenacity or Cohesion
Thermal Changes
Thermal Conductivity
Toxicity
Uniform Replacement
Vapor Pressure
Viscosity

AVAILABILITY

Unavailability is not a new phenomenon. Shortages or unavailability of certain materials have existed at certain times in various parts of the world—grain in Egypt during Biblical times; rice in China in modern times; oils and fats in Germany during the first World War; quinine, rubber, and other monopolistically controlled commodities because of restrictions in production and sale prior to World War II; silk, quinine, rubber, coconut oil, and many other raw materials due to crop failure, export or import prohibition, restrictions by producers or governments, unavailability because of rebellion or war.

Availability—the ability to get a material when and where it is wanted is of paramount importance. No matter how good a substitute may be, it is useless if not available. Therefore, the first step is to make sure that the substitutes to be examined are in plentiful supply, that they be preferably of domestic and not foreign origin, that the known suppliers will be able to take care of quantity requirements as needed.

With conditions being as they are today, it is seldom possible to plan ahead as far as raw materials are concerned. What is available today may soon become unavailable. However, since all consumers are subject to this same uncertainty, everyone is on an equal footing. That is why secondary substitutes must be decided on—in the event that the best substitute cannot be obtained.

HOMOGENEITY

Homogeneity implies uniform composition so that every part of a sub-stance or mixture is of identical composition and appearance. Some prod-ucts change in homogeneity because of differences in specific gravity, solu-bility, or for other reasons and produce a nonuniform condition. They must either be stabilized against such change or must be mixed or warmed to produce homogeneity before use. If the material is such that it cannot be brought back to its original state of uniformity (e.g., a decomposed glue), then it should not be used.

UNIFORM REPLACEMENT

Uniform replacement refers to the ability to obtain the same grade of product each time that it is ordered. Slight variations when unavoidable, may be compensated for, by technical control. Large or certain types of variations may make a product unusable. Thus, traces of copper are unde-sirable in materials used in rubber compounding. Consequently a material, which is the same in all respects, as previous deliveries, would be ruled out if contaminated with copper or its compounds. A material containing 0.1% of iron might be suitable for another purpose, but if the iron content in-creased to 0.5%, it may no longer be usable.

GRADE OR PURITY

The following grades of chemicals may be available:

C.P.
Commercial
Technical
Special
Natural
Synthetic
U.S.P.
Ultrapure
B.P.
Unofficial
N.F.
N.N.R.

C.P. stands for chemically pure. Each container usually bears a label of analysis, indicating the amounts of impurities present. This grade is usually the purest grade of chemical available. It is generally more expensive than the other grades but it is not specified except when high purity is required.

Commercial is the most common grade of chemical sold. Any chemical which does not bear a grade designation can be considered of commercial grade.

Technical is the ordinary commercial grade or some slight variation from it. This grade should not be used for food, drug, or cosmetic purposes without investigation.

Special refers to a particular grade made for a particular consumer or industry. It is different, in degree, from all other grades—either more or less pure. Its form and packaging may also be different.

Natural refers to a crude or refined product of vegetable, mineral, or animal origin; e.g., crude or resublimed iodine or camphor.

Synthetic refers to a chemical which is built up from a number of different chemicals by a chemical reaction process, e.g., synthetic menthol. A synthetic chemical usually contains fewer impurities than the corresponding natural product.

U.S.P. refers to the United States Pharmacopoeia, an official compendium giving the requirement for purity for many drugs and chemicals. A drug or chemical marked U.S.P. indicates that it meets all the specifications of the U.S. Pharmacopoeia.

Ultrapure refers to a most highly pure product.

B.P. refers to the British Pharmacopoeia, an official British compendium giving the requirements for purity for many drugs and chemicals. A drug or chemical marked B.P. indicates that it meets all the specifications of the British Pharmacopoeia.

Unofficial indicates that the drug and chemical has not been tested by the proper authorities and that it is not yet officially recognized in the pharmacopoeia.

N.F. shows that the product is listed in the National Formulary and is recognized by the American Pharmaceutical Association.

N.N.R. shows that the product is listed in New and Non-Official Remedies and is recognized by the American Medical Association.

FORM

Materials occur or are produced as gases, liquids, or solids. These are always the same under the same conditions of temperature and pressure. Gases and liquids usually do not exhibit any variation in appearance, handling, or use under similar conditions. Solids, however, do differ and may cause trouble. If they are crystals, the crystals may be large or small. A substitute may have a different crystalline form or shape (needle-like, cubic, etc.) which if used dry, may be undesirable because of appearance or bulking properties. Powders, likewise, consist of particles which may vary in size. Such variations not only affect appearance but also density, flow, agglomerating or "caking" tendencies, suspension, deposition, friction, and other properties.

OPTICAL PROPERTIES

Color (Shade, Intensity)
Clarity
Fluorescence
Phosphorescence
Iridescence (Pearliness)
Refractive Index
Reflectance (Dully, Shiny)

Color is of importance not only for appearance but also where staining, dyeing, or pigmentation occurs. The color of a material may vary with the size of the particles, larger particles being darker. Thus, crystalline copper sulfate is blue while the finely powdered material is a very light blue. Certain materials lose or change their color on being dissolved, dehydrated, or on interaction with another ingredient.

Clarity refers to clearness and freedom from haze or turbidity. Most commercial products are clear. Sometimes they develop a haze, turbidity, deposit, or a sediment, especially in metal containers. Others lose clarity even in glass containers because of polymerization (e.g., formaldehyde).

Fluorescence is the instantaneous re-emission of light from a substance of a greater wavelength than that of light originally absorbed. Common examples are seen in a solution of fluorescein in water and in certain types of mineral oils.

Phosphorescence is the re-emission of light, after a time lag, of a longer wavelength than that absorbed. This phenomenon is typified by the glowing of yellow phosphorus in the dark.

Iridescence is the rainbow-like play of colors as of pearls and soap bubbles.

Refractive Index is the relationship between the speed of light in a vacuum and its speed in a substance. The refractive index of a substance determines the degree of bending or distortion of an object viewed through the substance. Thus, it is of importance in adhesives for cementing optical glass, transparent plastics, etc.

Reflectance refers to the fraction of light which is reflected when light falls on any surface. Thus, a rough surface reflects very little whereas a smooth surface reflects more light. The former appears dull and the latter, shiny.

ODOR

Pleasant
Unpleasant
Strong
Faint
Temporary
Permanent

Odor is the effect on the sense of smell produced by particles emanating from a substance. In many products such as food, cosmetics, and household articles, odor is an important factor. Where an undesirable odor cannot be eliminated, it may often be "covered up" by a stronger, more desirable odor.

No odor is equally pleasing to all. Certain types of pleasant odors are bland, refreshing, or stimulating and are not objectionable in certain products. Unpleasant odors may be sickening, irritating, or depressing. An odor may be strong or faint. Faint, unpleasant odors are more tolerable than strong, unpleasant odors and may be masked more easily.

Very volatile odors may only be temporary and may disappear quickly on aging, storage, or use. Permanent odors must be recognized as an everpresent factor.

In blending various materials there may be a diminution of odor caused by the dilution or change in character or strength of the substance. These changes may result from decomposition or interaction with another ingredient.

TASTE

Sweet
Sour
Bitter
Salty
Spicy
Oily
Fruity
Neutral or Tasteless
Pleasant
Unpleasant
Strong
Permanent

Taste is a factor in those products that enter the mouth. Such products are foods, beverages, medicines, dentifrices, and certain cosmetics for the lips.

Pleasant tastes may be sweet (as in sweet chocolate); sour (as in lemon drops); bitter (as in hops, used in beer-making); spicy (as in ginger); salty (as in brine); oily (as in olive oil); neutral or tasteless (as in water); fruity (as in berries).

Just as with odors, strength and permanence are of importance and must be given due regard. An undesirable taste may often be "covered up" by a stronger or more desirable taste. Certain tastes which are unpleasant when too strong, are more pleasant when diluted, e.g., saccharine.

pH

pH is the logarithm of the reciprocal of the hydrogen ion concentration in gram molecules per liter or, more simply, a measure of acidity or alkalinity of a water solution of a substance. Pure water, which is neutral, has a pH of 7. Any pH above 7 is considered alkaline and below 7 is considered acid.

Thus, the pH of a solution of a material is indicative as to whether it is alkaline or acid and sometimes is a measure of its strength. This is a clue to how it will affect materials with which it is mixed or with which it may come into contact. Further details of the influence of acidity and alkalinity are given in the section on *Interaction With Other Materials*.

DENSITY AND SPECIFIC GRAVITY

Density is the weight per unit volume; e.g., pounds per cubic foot. Specific gravity is the relation between the weight of a given substance compared with the weight of an equal volume of water at the same temperature and pressure.

The density or specific gravity of a product varies with its purity, porosity, size of its particles, and the process by which it was made.

Density or specific gravity are critical factors where bulking, value, suspension, low cost, etc. are important.

Thus, calcium carbonate will vary in density or specific gravity depending on whether it is in the form of natural limestone, marble, or chalk, or a chemically precipitated product.

VISCOSITY

Viscosity is the resistance of a fluid to shear, agitation, or flow. More commonly it refers to rate of flow of a specific liquid as compared to water or any other commonly used liquid.

In some cases, viscosity is of importance because the greater the viscosity of a liquid the lower the rate of flow, spreading, penetration, wetting, etc., and the better its suspending power. A lower viscosity of course reverses these properties. A viscous liquid is harder to mix, fill, pour, and apply than a less viscous liquid.

Viscosity may be increased or lowered by suitable additions and treatments. Thus, the viscosity of mineral oil can be increased by heating it with some aluminum stearate; the viscosity of an alkaline casein dispersion can be reduced by means of urea. Other specific methods for altering viscosity are known and these should be used when a substitute is suitable in all other respects.

GELLING OR THIXOTROPIC TENDENCIES

Gelling is the formation of a gel or jelly-like substance; e.g., glue or agar with water. The thixotropic state refers to a gel which liquefies on shaking or stirring and which regels on standing; e.g., iron hydroxide or certain clay suspensions in water.

Gelling may be desired in certain cases as in hectograph (duplicating) compositions, whereas in the case of a paint, gelling, which would prevent brushing or spraying, is undesirable.

Gelling may be due to the colloidal properties of a single substance in a liquid (as with gelatin and water) or may result from the interaction of one or more substances (as with sodium silicate and dilute hydrochloric acid).

Gelling may be temporary, as in the case of a cold gelatin and water jelly, which becomes liquid on warming; or it may be more or less permanent as in the case of rubber cement (rubber swollen in a hydrocarbon solvent).

Gels may be thinned or prevented from forming by the addition of suitable agents. Thus, fish glue in water is prevented from gelling by the addition of acetic acid.

FREEZING POINT (MELTING POINT)

The freezing point is the temperature at which a liquid solidifies or begins to form crystals under normal conditions. Liquids containing impurities or added substances have different freezing points than the pure liquids. Therefore, the freezing point of a liquid is a measure of its purity. Similarly, if the freezing point of a substance is too high or too low, it may be altered by suitable additions.

The melting point is that temperature at which a solid changes to a liquid under normal conditions. The melting and freezing point of any substance is usually the same.

Some substances (mixtures), e.g., hydrogenated coconut oil, do not have a definite melting point but melt over a specific temperature range. Other substances soften or become plastic at certain temperatures; e.g., pitch, cellulose acetate, etc. Still others do not melt but sublime when heated sufficiently.

VAPOR PRESSURE

Vapor pressure is the pressure of any vapor above its liquid or solid form at the temperature at which equilibrium is established.

The greater the vapor pressure of a substance, the greater is its tendency to evaporate and disappear when exposed. High vapor pressure is desired in products which are expected to evaporate or dry quickly, as in cleaning fluids and lacquer thinners. Low vapor pressures are desired in products which should not change in bulk or dry out as in flexibilizers for glue, casein, etc., or plasticizers for lacquers or plastics.

SUBLIMATION

Sublimation is the direct vaporization of a solid that does not first liquefy; e.g., camphor or naphthalene.

Substances that sublime are useful when volatilization at certain temperatures is desired. Certain substances (camphor and naphthalene) sublime at ordinary temperatures. Of course, this means that the latter gradually disappears when exposed. Where such volatilization is undesirable, subliming substances should not be used.

BOILING POINT

The boiling point is the temperature at which the vapor pressure of a liquid equals the atmospheric pressure. Pure liquids have a definite boiling point. Commercial products, which contain impurities, boil over a range of temperatures, known as the boiling range. Thus, pure water boils at 100 C at 760 mm pressure. Commercial methyl oleate boils at 200–215 C at 15 mm pressure.

Low boiling liquids volatilize readily and disappear. This, of course, is advantageous where quick drying is necessary, as in the case of rubber cement or hair lacquers. High boiling liquids are specified where volatility is to be kept at a minimum to prevent drying out, brittleness, shrinkage, etc., as in the use of glycerin in Cellophane or castor oil in ethyl cellulose.

Boiling points can be varied by dissolving soluble materials in a liquid or by mixing it with another liquid of a different boiling point. In the former instance, the boiling point is raised whereas in the latter it is either lowered or raised, depending on the boiling point and solubility of the added liquid.

Certain mixtures (azeotropic) form a constant boiling mixture, at a temperature different than that of any of their components. Here the distillate has the same composition as the substance being distilled.

SOLUBILITY

Solubility is the weight of a substance that can be dissolved in a definite weight of solvent at a given temperature. Thus, 100 g water dissolves 35.8 g salt at 10 C.

Most substances are more soluble in hot water or other solvent than in the cold. Therefore, it is important to determine the solubility of a substance in the solvent in which it will be used, in the temperature ranges to which it will be exposed. Otherwise precipitation may result on later cooling or heating due to temperature changes.

Many substances form a true solution when added to a solvent; e.g., salt in water or menthol in alcohol. Other substances disperse colloidally, as does soap in water or nitrocellulose in acetone. Colloidal substances usually swell slowly and a new material, if it is colloidal, should not be hastily discarded on account of poor solubility. Let it stand overnight, with the solvent, before stirring.

Certain substances may not dissolve in either of two solvents, but will dissolve in a suitable mixture of the two; e.g., nitrocellulose in ether and alcohol.

HARDNESS

Hardness can be measured by the ability of a substance to abrade or scratch other materials or conversely the ability to be abraded or scratched by other materials; or by the depth of penetration of a sharp edge or point under a definite weight or pressure.

Thus, if a coating is to be applied to a floor, it should be sufficiently hard to withstand a certain amount of wear. For this reason, in a wax polish, carnauba wax, which is harder than paraffin wax, is preferable.

An abrasive for rough polishing of quartz must be harder than quartz; e.g., corundum would be satisfactory.

A typical scale of abrasion hardness is that of Moh given below, with #1 as the softest and #10 as the hardest:

1. Talc
2. Gypsum
3. Calcite
4. Fluorite
5. Apatite
6. Orthoclase
7. Quartz
8. Topaz
9. Corundum
10. Diamond

Penetration tests of hardness made with a penetrometer or durometer are expressed, in the case of the former, in units referring to a standard material under certain conditions and, in the latter, in units on a calibrated dial. Rockwell and Brinell hardness values are used for referring to the hardness of metals, alloys, and similarly hard materials.

TENACITY OR COHESION

Tenacity or cohesion is the property preventing a substance from breaking into pieces when struck, pressed, or pulled strongly.

Certain materials possess tenacity or cohesion inherently, as rubber or sisal fiber (fresh). Others require the addition of a plasticizer; e.g., as in "Pliofilm," Cellophane, etc. This property varies in degree in all materials and also varies in accordance with the ingredients with which it is mixed and the process of manufacture. Thus, unvulcanized or soft vulcanized rubber is tenacious and coherent, whereas highly vulcanized hard rubber is brittle.

PLASTICITY AND DUCTILITY

Plasticity is that property of a material that permits the altering of its shape or size by the use of pressure, tension, heat, or a combination of these forces, and does not permit a return to the original shape with the removal of these forces. Examples of plasticity are seen in butter and gelled glue solutions. When heat is one of the elements producing plasticity in a substance, the substance is said to be thermoplastic. When the shape or size becomes fixed and cannot again be altered by the original or any other forces, the substance is said to be thermosetting; e.g., Bakelite. When maintenance of shape and size is important, the use of plastic deformable materials should be avoided; whereas, if rigidity is not desired and alterations in shape or size are immaterial or quite desirable, a plastic deformable body should be used.

Ductility is that property which permits the drawing out of a body in the direction of its length; i.e., into a continuous thread, wire, tube, or rod; e.g., rayon thread, steel wire, lead tubing, or brass rod. Thus, when continuous threads, wires, rods, or tubes are to be made by the drawing out of a material, it is important to use only those materials which exhibit a high degree of ductility.

FLEXIBILITY

Flexibility refers to the ability to bend repeatedly within limits under certain conditions without cracking or breaking. Thus, soft rubber qualifies as a flexible material but hard rubber is inflexible and brittle. Most materials are more flexible at higher than at lower temperatures. The addition of suitable plasticizers or flexibilizers increases flexibility in many commercial products; e.g., dibutyl phthalate in lacquers and certain plastics.

ELASTICITY

Elasticity is the property of recovering original shape and dimensions after stretching, squeezing, or twisting. Soft rubber is the classic example of an elastic material. Elasticity of many substances varies with age, use, and admixture with other ingredients.

LENGTH

Length refers to the stringiness of a fluid when it is poured or when a rod is dipped into it and then pulled out. For example, mineral oil possesses length after it has been heated with a certain amount of aluminum stearate; also a strong "solution" of gum karaya in water. Length is required in certain products to decrease their rate of flow and penetration so that they will remain in place, for a longer time.

STICKINESS

Stickiness is the property of a substance to adhere to anything that it touches. It is advantageous in adhesives, glues, cements, etc., but is detrimental to many other uses. Certain products require adhesiveness on one side and nonadhesiveness on the other. Thus a paint should stick to the surface to which it is applied and dry, free from adhesiveness, on its exposed surface.

In some cases stickiness must be permanent, as in fly paper. In other cases it must be temporary as in rubber cement for paper. In the latter case, it eventually dries (without losing its adhesiveness) and can be peeled off without damaging the paper.

Adhesiveness may be increased by suitable additions or processing, for example, the addition of an alkaline casein solution to rubber latex. Conversely, stickiness may be decreased, as in the case of the addition of "Acrawax C" to the polyvinylbutyral resin that is used to replace rubber for coating raincoats.

ADHESION

Adhesion is the attraction that causes two surfaces to stick to each other so that they cannot be separated easily. This is usually brought about by means of an agent which solidifies by drying, cooling, chemical or physical change. Adhesion differs from stickiness (previously mentioned) which produces a bond between two surfaces which can be easily separated by pulling.

Adhesion varies with the composition of the surfaces to be united, their smoothness or roughness, and other factors. The choice of an adhesive material is dependent on whether it is to be applied hot or cold, wet or dry, in aqueous or other solution; whether it is to "set" slowly or quickly; whether it is to be temporary or permanent. The choice is also dependent on conditions of temperature; stress; contact with water, solvents, or chemical; and effect on the materials with which it is used; etc.

SLIPPERINESS

Slipperiness is the property of a substance to slide or move with very little friction.

Certain materials such as oils, greases, waxes, molybdenum disulfide, and graphite are inherently slippery. Of course they vary in degree and with the nature of the surfaces on which they are used. Other pertinent factors are temperature and load.

Slipperiness can be increased by the addition of ingredients mentioned above. Conversely, slipperiness can be decreased by the addition of adhesives; e.g., rubber or sand, respectively. Slipperiness may be temporary as in the case of linseed oil which dries to a nonslippery solid, or permanent as in the case of talc which is not readily affected by age or oxidation. Suitable additives are available for increasing the useful life of a slippery material.

DRYING QUALITIES

Most coating materials such as paints, lacquers, varnishes, etc. must dry fairly rapidly so as not to mar, stick, or hold dust, insects, etc. Rapid air-drying (at normal temperatures) is a prerequisite for them. Other products such as lubricants or flypaper must be nondrying.

Drying may be due to evaporation, absorption, oxidation, polymerization, or other factors. Time of drying may be modified by admixture with other substances, change of temperature or pressure, chemical treatment, contact with catalysts or inhibitors, etc.

FEEL OR "HAND"

Feel or "hand" refers to the sensation felt by the fingers or other parts of the body in contact with a material. Thus, jute is rough and rayon is smooth. Wood, being a poor conductor of heat, is warm to the touch whereas aluminum, a good conductor of heat, is cool to the touch.

Rough materials can be made smooth by mechanical processes such as grinding, polishing, weaving, compressing, or coating with wax, resins, or starch. Smooth materials may be made rough by grinding, sand blasting, garnetting, and other mechanical processes as well as by treatment with certain chemicals. Materials that feel warm can be modified by combining with good conductors of heat, as by weaving a fabric from textile and metal threads. Conversely, a metal can be made to feel warmer by mixing the powdered metal with a plastic or fiber powder and compressing them until they form a homogeneous body.

CRYOSCOPIC PROPERTIES

Cryoscopic properties refers to the ability of a substance to lower the freezing point of a liquid. Thus, water which ordinarily freezes at 32 F is protected against freezing to about 20 F by the addition of 9 pints of 60% glycerin per gallon of water. Most chemicals that dissolve in water or other solvent, affect its freezing point. Caution should be observed therefore in making substitutions, in preparations that may be exposed to low temperatures, as freezing will not only cause congealing or crystallization but may cause expansion that might rupture the container.

In many cases a frozen product reverts to its original state of uniformity when it thaws out. In other cases, as in certain emulsions, particularly those made with vegetable gums, freezing breaks the emulsion and the latter does not re-emulsify on thawing out.

HYGROSCOPICITY OR EFFLORESCENCE

A material is said to be hygroscopic when it can absorb moisture from its surrounding medium, usually air; efflorescent when it loses moisture when exposed to dry air.

While hygroscopicity is often useful in preventing drying out or embrittlement, it may be undesirable where caking or lumping of dry materials or liquefaction results, because of it.

Efflorescent materials tend to give up moisture and dry out and lose their original bulk or crystal form.

Both of these properties can often be neutralized by coating dry materials with gums, waxes, fats, oils, or other film forming materials which prevent the ingress or egress of moisture.

INFLAMMABILITY

The flash point of a substance is a measure of its inflammability. Specifically, it is the lowest temperature at which a substance, in an open vessel, gives off enough combustible vapors to produce a momentary flash of fire when a small flame is passed near its surface.

Flash point must be considered not only from the standpoint of a fire hazard, but because of local, state, and national regulations for storing, shipping, and using products that are inflammable. Thus, a "hiflash" naphtha is less hazardous than a lower boiling naphtha or gasoline.

Burning point is the lowest temperature at which a substance will burn when a source of heat is applied under specific conditions. Thus, different woods will burn at different temperatures, depending on their nature, dryness, etc.

EXPLOSIVENESS

Explosiveness refers to the tendency of a material or its combustion or decomposition products to burst or expand violently. An explosion is usually initiated by impact, heat, pressure, an electric spark, contact with air or another substance, or a catalyst.

Certain explosive materials are quite safe if carefully handled, whereas others are very sensitive and extremely dangerous. An example of the former is dynamite and of the latter is mercury fulminate.

The choice of a substitute explosive material should involve consideration of its condition of handling, storage, use, and transportation.

TOXICITY

Toxicity refers to harmful physiological effects of a substance on living beings and things. The toxicity of many materials must not only be considered because of their effect on human beings but also on animals, fish, and plants, with which they may come in contact. Nearly every substance is toxic if taken in or applied beyond a certain limit. For example, salt is not harmful in small quantities. Larger amounts will produce vomiting and other harmful effects. Similarly, lemon oil which is used in baking candy, and other internal preparations becomes very toxic and corrosive when used in larger doses.

Some substances exert a toxic effect through their vapors (particularly in confined spaces) as mercury; others do when taken through the mouth as sodium fluoride; and others, by absorption through the skin, as p-toluidine.

By providing proper conditions and utilizing all safety measures, many toxic substances may be used with a minimum of hazard.

EFFECT ON SKIN, HAIR, OR FINGERNAILS

Some materials produce undesirable effects on the skin, hair, or fingernails. These effects, especially in small amounts, may not be dangerous but they are nevertheless undesirable.

As typical examples, cresols irritate and cause the skin to crack; sulfuric acid corrodes the skin; picric acid stains the skin; acetone and other solvents cause brittleness of fingernails; alkali (lye) causes roughening and reddening of the skin and hair; alum or formaldehyde causes toughening of the skin; vegetable oils cause softening of the skin, nails, and hair.

All of these effects are dependent on concentration of the substance and time of contact. Therefore, if a substance is present in only certain small amounts and in contact but momentarily, it is not necessarily objectionable. But substances of this nature should be carefully evaluated before use.

EDIBILITY

A substance is considered edible if it possesses food value and is not harmful when eaten in certain amounts. Thus, the use of glycerin cakes is permitted because it has food value and is not harmful in the amounts used. Mineral oil, while not harmful, is not permitted in mayonnaise because it has no food value. Physiological data should be obtained from the supplier or from authoritative tests, before using a new material for edible purposes.

THERMAL CHANGES

Temperature can produce many effects on different materials. Heat may change a liquid to a gas (ether); a solid to a vapor (naphthalene); a solid to a liquid (paraffin wax). Cold, of course, reverses the above changes. Crystalline materials may melt and fuse to a solid mass and not revert to crystals when cooled (abietic acid crystals). Solid objects may flow out of shape on heating with a resultant change in appearance (e.g., thermoplastics, like cellulose acetate).

Most materials shrink on cooling but water expands on being frozen. Most materials expand on heating but ice shrinks on being melted. Each material shows a definite amount of expansion or shrinkage on heating or cooling. This information may be of importance in storage and use of certain products.

Heat decomposes or alters some substances (egg whites). The effect of heat on color, odor, taste, viscosity, and all other properties mentioned in this chapter should be noted.

Cooling often causes precipitation or thickening. Freezing "breaks" emulsions, which, in many cases, do not revert to their original form on thawing.

Heating may also produce changes in stickiness, adhesion, optical properties, pH, odor, taste, density, viscosity, "length," vapor pressure, solubility, hardness, flexibility, tenacity or cohesion, feel, slipperiness, elasticity, hygroscopicity, inflammability, explosiveness, drying qualities, plasticizing properties, chemical reactivity, stability, toxicity, effect on skin, hair, or fingernails, edibility, pressure, solvency, bacterial content, emulsifiability, surface tension, wetting-out properties, dispersing properties, adsorption and absorption, electrical properties, radioactivity, etc. It would be prudent, therefore, to refer to each of the above named headings and try to judge how temperature changes in any substitute may affect its final use.

Cooling usually produces an opposite effect to that of heating.

EFFECTS OF PRESSURE

Increased pressure usually increases the solubility of gases in liquids (carbon dioxide in beverages). It may also increase the speed of chemical reaction. It usually increases density, viscosity, hardness, and rigidity. Conversely, it decreases vapor pressure, flexibility, elasticity, absorption, and size.

SOLVENCY

Solvency refers to the ability of one material to dissolve another at a given temperature. For example, the solvency of water for borax is 14.2 g per 100 cc at 55 C. One ingredient may be used primarily as a thickener or to prevent drying out. If its solvent properties are ignored, precipitation or crystallization may result. Thus, corn syrup may replace glycerin as a thickener but its solvent powers for vanillin, borax, and many other ingredients are much lower, thereby causing possible difficulties.

PLASTICIZING OR FLEXIBILIZING PROPERTIES

A plasticizer or flexibilizer is used because of its ability to prevent brittleness or cracking of a solid material; e.g., glycerin with glue or dibutyl phthalate with pyroxylin.

A good plasticizer should be compatible with the material in which it is used and be sufficiently nonvolatile to remain in it for its useful life. It should not change appreciably the properties desired in the finished product. The amount of plasticizer used is often critical—too much gives a soft or sticky product, too little is not sufficient to plasticize properly.

INTERACTION WITH OTHER MATERIALS

Chemical reaction between two ingredients may produce changes which rule out the finished product. Therefore, it is best to avoid the substitution of a substance which may react with any of the substances already present.

Acids may produce esters or salts or induce hydrolysis. Thus, adding acetic acid to alcohol will produce some ethyl acetate on standing. The addition of phosphoric acid to a caustic soda solution forms sodium phosphate. Ethyl lactate will be hydrolyzed by the addition of an aqueous acid, such as hydrochloric acid.

Alkalies may form soaps, induce hydrolysis, or form hydroxide gels or precipitates. Thus, the addition of caustic potash to a solution of castor oil in alcohol produces a castor oil soap. An ester such as methyl oleate will be hydrolyzed by an alkali such as sodium carbonate. The addition of ammonium hydroxide to a solution of ferric chloride produces a colloidal precipitate or gel of iron hydroxide.

Salts may produce electrolytic effects such as increased conductivity of an electric current or generation of an e.m.f. which may induce corrosion. Distilled water is a poor conductor of electricity but the introduction of an electrolytic salt such as sodium chloride, increases its conductivity greatly.

Oxidation or reduction of other ingredients or of the newly added material by the other ingredients may produce many changes in properties.

Hydrogen peroxide oxidizes many organic materials with a lightening or darkening effect. Reduction with hydrogen (produced from metal and acid) also often brings about undesirable changes.

Double decomposition of two substances with resultant precipitation or formation of new gaseous or liquid substances may result in certain cases.

Thus, if barium chloride is added to a solution containing a sulfate such as sodium sulfate, a heavy, white precipitate of barium sulfate forms. If sodium acid sulfate is added to sodium bicarbonate, a gas, carbon dioxide, is formed.

Bacterial or enzymatic reactions may be induced by the new material or it may itself be thus affected. Many organic substances such as starch, sugar, etc. are decomposed by bacteria or enzymes. Such decomposition can be avoided by sterilization or by the use of suitable preservatives.

The addition of a substance having an absorptive or adsorptive effect may abstract an amount of material so as to change many characteristics of the composition. Thus, adding bentonite to a gum tragacanth oil and water emulsion, abstracts water and breaks or thickens the emulsion considerably.

STABILITY

Stability refers to the property of a substance to remain substantially unaltered over a certain period of time. A substance or composition is subjected to so many detrimental conditions, that it is really surprising that commercial products which remain in a dealer's stock for extended periods, are still suitable for use when purchased. This is not due to luck but to judicious testing and aging.

Water, when present, is useful in many products but detrimental to others. Thus, a dry starch preparation keeps indefinitely. The introduction of even a small amount of water may cause bacterial decomposition which will ruin the product.

Air, which is so necessary to living things, is detrimental to many substances. Thus, cottonseed oil tends to absorb and react with the oxygen in the air and becomes rancid.

The effects of heat and cold have been indicated under *Thermal Changes.*

Sunlight may improve some products (irradiated foods) and injure others. An example of injury by sunlight is the effect of the latter on rubber, which becomes brittle and cracks.

Motion often produces detrimental effects. A powdered mixture consisting of materials of different densities (bentonite, soda ash, and soap) may separate into different layers, thus destroying uniformity because of the motion produced during transportation.

Age may also produce various changes. Certain crystalline rearrangements take place in metallic alloys on aging. Unless these are properly controlled, the alloy may be unserviceable.

The electric discharge or "corona" effect producing ozone is detrimental to many organic products. Thus, rubber insulated electrical ignition wires deteriorate rapidly when exposed to electrical discharges.

Abrasion is a factor that must be considered when selecting a material that is subject to friction or pounding. Thus, while marble makes beautiful steps, it wears down much more rapidly than a harder material like granite.

The material in which a composition is to be packaged is of utmost importance. Thus, it would be foolish to package a caustic paste in aluminum tubes as the latter would be readily attacked and soon destroyed. Even ordinary water solutions corrode beneath the tin-plate in cans. To avoid these corrosion effects, either other suitable packaging materials are selected, or, when possible, suitable inhibitors are added to stop or delay such action.

BACTERIAL CONTENT

While a material may not be susceptible to bacterial decomposition, it may act as a carrier for bacteria which may affect other ingredients. Thus, with compositions containing organic matter on which bacteria may thrive, it is necessary to avoid the introduction of bacteria or to destroy them by sterilization or antiseptic action.

EMULSIFIABILITY

Emulsifiability is the ability to form an emulsion with one or more immiscible fluids. Thus, cottonseed oil is emulsifiable with water and gum tragacanth. The latter is the emulsifier. Different substances have different degrees of emulsifiability and therefore require different types and amounts of emulsifiers as well as different processing methods.

Vegetable oils, for example, can usually be emulsified with ammonium linoleate. Paraffin wax, in the presence of salts, is usually emulsified with gelatin or gum arabic. Toluol with acetic or other acids requires a special emulsifier that acts in the presence of acid such as "Emulgor A."

For certain purposes, a water-in-oil or an oil-in-water type of emulsion is desired. The introduction of a substitute ingredient may produce a type of emulsion opposite to the one desired. Sometimes a change in type can be made by changing the emulsifier or the proportions of one or more ingredients or by reversing their order of introduction and mixing or by altering the pH.

Variations in quantity and identity of ingredients and methods of emulsification also produces variations in particle size, color, viscosity, and other properties.

SURFACE TENSION

Surface tension refers to the tendency of the surface of a liquid to contract to the smallest area possible. The lower the surface tension, the greater the contraction. A high surface tension produces the opposite effect.

Low surface tension permits rapid and thorough wetting of an insoluble material. Thus, the addition of a small percentage of a wetting agent like a sodium salt of a sulfonated hydrocarbon to a water solution of a flame proofing agent like sodium borophosphate, permits a fabric to be impregnated quickly and thoroughly.

The addition of a wetting agent to various solutions makes them spread more quickly and evenly on smooth surfaces such as glass, steel, etc.

Certain oils are caused to penetrate more quickly and deeply into crevices when a suitable surface tension reducing agent is added to them.

DISPERSING PROPERTIES

Dispersing properties refers to the ability of a substance to suspend insoluble particles in a fluid. For example, colloidal carbon is suspended in water with gum arabic to form India ink. Pigments are suspended in drying oils by suitable grinding with a polyglycol laurate.

Proper dispersion of insoluble particles in a liquid is important to avoid deflocculation or settling and in forming uniform films and coatings as in paints and pigmented lacquers.

ADSORPTION AND ABSORPTION

Adsorption is the taking up (concentration) of a substance on the surface of another substance. This is illustrated by charcoal taking up odors in a refrigerator. Absorption is the taking up of a gas or vapor, by a fluid in which it dissolves; e.g., the dissolving of acetylene in acetone.

Adsorptive materials may take up certain materials and thus throw a formulation out of balance. They may also remove or lighten the color or odor of a product and thus produce an unwanted change.

HEATING POWER

Heating power refers to the property of a substance to increase the temperature of another substance. Thus, the substitution of butane (which has a different caloric value) for gasoline fuel, produces a different amount of heat. The addition of caustic soda solution to aluminum produces an exothermic reaction generating much heat. Simple solutions of many materials in water produce a heating or cooling effect. (See *Thermal Changes* for changes in properties produced by heating and cooling.)

LIGHTING POWER (CANDLE POWER)

Lighting power is the ability of a substance to produce illumination. Materials such as illuminating gas, kerosene, etc., are used for producing illumination. Every pure hydrocarbon produces a definite candle-power when it burns under certain conditions.

Metals such as tungsten or zirconium alloys produce light when heated sufficiently; e.g., an electric current. Carbon and other conductive refractories produce light when an electric arc is formed between two electrodes in an electrical circuit.

Gases, subjected to an electric discharge are also used to produce light, as in neon tubes.

ELECTRICAL PROPERTIES

Conductivity, the reciprocal of resistance, is the rate or degree of transmission of electricity through a substance. Thus, silver and copper are excellent conductors of electricity and rubber and slate are poor conductors (insulators).

If any other electric properties or effects are of importance, the necessary data should be gotten or specific tests made as outlined in electrical texts or handbooks.

THERMAL CONDUCTIVITY

Thermal conductivity refers to the degree or rate of heat transmission through a substance. Good conductors of heat are exemplified by silver and aluminum. Poor conductors of heat (insulators) are typified by asbestos and "Fiberglas."

RADIOACTIVITY

Radioactivity is the property of a substance to emit rays that can penetrate many solids, ionize air, and excite phosphorescence in certain substances.

Radium is typical of radioactive materials and use is made of it in taking photographs of ordinary opaque objects (e.g., the human body). Uranium compounds are used in luminous paints for watch and instrument dials.

Radioactive materials differ in intensity and in span of life. They are also subject to change in activity by admixture with other substances.

POLYMERIZATION

Polymerization can cause changes in many physical and chemical properties.

HANDLING

The selection of a substitute may introduce new problems in handling. Often this is of minor importance. Sometimes it may require new or changed equipment and special precautions.

Thus, when a solid material is to be replaced by a liquid, it may have to be stored in tanks instead of bins and pumped instead of dumped or shoveled.

In changing from the use of a solid to a liquid (or a gas), or vice versa, all the headings of this chapter should be scrutinized for possible sources of difference in receiving, storing, using, grinding, mixing, melting, dissolving, etc.

LEGAL RESTRICTIONS

The purchase, ownership, storage, use, or sale of many materials and products is covered and bound by national, state, city, or local regulations.

Foods, drugs, and cosmetics must conform to the regulations of the U.S. Food and Drug Administration and State and City Departments of Health.

Habit-forming drugs such as cocaine and morphine are strictly regulated by the federal government under the Narcotics Act.

Certain states and cities require licenses for the manufacture and sale of drugs, cosmetics, and other products.

Explosive regulations of the U.S. Bureau of Mines require a license for owning any type of explosive material even though it is to be used in a nonexplosive product.

The use of new materials is not only subject to the above regulation (which is far from complete in scope) but to State and local laws on production of disagreeable odors, toxic materials, and other hazardous products that may increase the fire or explosion hazard or injure anyone in or near the factory in question.

Since many substances and their uses are patented it is also well to ascertain whether the use of a substitute (for commercial purposes) will mean infringement on a patent covering the same.

Miscellaneous

FORMULATION

The use of a substitute material may involve only a slight or a radical change in formulation. If only a change in proportions (see *Proportions*) is necessary, then reformulation is relatively simple; e.g., the replacement of glycerin by ethylene glycol in an antifreeze mixture for automobile cooling systems.

A more complicated case is the use of polyvinyl alcohol as an emulsifying agent for a carnauba wax water emulsion. There are many emulsifying agents such as ammonium linoleate or other fatty acid derivatives which can function equally as well as polyvinyl alcohol as an emulsifying agent, but they are not suitable because they are not film-forming adhesives. The latter quality is necessary in a particular coating problem. The nearest logical substitute would be a film-forming adhesive type of emulsifier such as methyl cellulose. In doing this, different proportions of emulsifier wax and water, mixed under different conditions would have to be tried to get optimum results. It might be necessary to add a secondary emulsifier or stabilizer and perhaps even a plasticizer or solvent coupling agent such as "Carbitol."

The replacement of dibutyl phthalate as a plasticizer in lacquers is also an example of a more complicated case. When diethylene glycol monolaurate is substituted for the former, it is necessary to use a smaller amount of the latter and also vary the resin and solvent proportions in order to get a good lacquer.

Sometimes, the use of a substitute may entirely change the character and use of a product. For example, trichlorethylene, a low boiling liquid (non-aqueous) used in the degreasing of metals may be replaced by sodium abietate, a dry crystalline material which is used with water. Here is an example of a substitute being totally different in its properties and method of use but serving the same function.

PROPORTIONS

It is seldom that one is fortunate enough to get a substitute for use in a composition that will replace the original material pound for pound. It may be necessary to use a greater or smaller amount of a new material to produce the desired result.

Thus, where a soap was formerly made from coconut oil and alkali and the former is to be replaced by castor oil, a larger amount of castor oil would be needed for complete saponification. This is due to the difference in combining weights of these two oils. Combining weights are only of importance in determining amounts to be used when a chemical reaction takes place.

Where a substitute material is to be used in the same volume as the material which it is to replace, then the finished product will usually have a different unit weight. Thus, in a composition containing 2 quarts of glycerin per gallon of solution, wherein the glycerin is to be replaced by 2 quarts of propylene glycol, the difference in weight per gallon of solution produced by this substitution is appreciable.

Certain materials that have great strength or activity are used in smaller amounts when substituting for weaker materials. For example, if hydrochloric acid is used in place of acetic acid, a much smaller amount will be needed. Similarly, caustic soda, which is a much stronger base than triethanolamine, would be used in a smaller amount. Conversely, when a weak material replaces a sgrong material a larger amount may be needed; e.g., replacing phosphoric by lactic acid.

METHODS OF MANUFACTURE

The different properties (as indicated in this chapter) of a substitute or alternative may present many problems in the manufacture of a finished product. In some cases this problem may be minor and in others more serious.

For example, if sugar is to be replaced by molasses, the latter may require a tank for storage; pumps and valves for delivering it to the mixing tank; exhaust fans for drawing off the odor; special materials of construction to prevent corrosion; cleaning and sterilization of all tanks, pipelines, valves, pumps, etc. to prevent fermentation. A sludge (due to impurities present in molasses, not in sugar) may also require special filtering apparatus or settling tanks. If color is a factor, it may be necessary to heat the molasses with a decolorizing carbon or other materials and then filter it.

In cold weather the storage tank may have to be heated to reduce viscosity and permit ready flow of the molasses. None of these factors is present when sugar is used because it is a dry powder or crystal material that can be dumped into the kettle from sacks or barrels and presents none of the problems mentioned above.

When a corrosive material replaces a noncorrosive material as in the case of calcium chloride in place of activated alumina for moisture absorption, the calcium chloride may corrode the metals formerly used and require the installation of materials at points of contact which are resistant to it.

COSTS

The relative price per pound of a substitute to that of the material which it replaces is not the sole economic criterion in making a choice. Thus, while propylene glycol is more expensive than glycerin, it may be more economical to use it in making an imitation vanilla flavor because of its far greater solvent power for vanillin. Accordingly, a much smaller amount of it is needed.

Sometimes a higher cost for a substitute may be justified if the finished product is superior in use or salability. For example, a stamp pad ink made with glycerin varied too much in consistency during dry and damp weather. Furthermore, the rubber stamps had a limited life period. Replacing the glycerin with glyceryl monoricinoleate produced an ink which varied very little with atmospheric moisture changes and prolonged the life of the rubber stamps considerably.

Substitutes which are usually introduced by necessity are sometimes continued in use after the necessity has ceased to exist. The reasons for this may be one or more of the following: ease in handling, uniformity, more than one source of supply, shorter manufacturing time, lower maintenance and labor costs, reduction in insurance, avoidance of license fees or patent suits, governmental regulations and recordkeeping, and other possible advantages.

USE OF THE FINISHED PRODUCT

The introduction of substitute or alternative materials may produce such changes in a product as to necessitate changes in the method of use by the consumer. This, while highly undesirable, is sometimes unavoidable.

For example, lemon extract as used by bakers, consists of a solution of lemon oil in alcohol. When the alcohol is replaced by emulsifying the lemon oil in water with a vegetable gum, a thick messy emulsion results, which is less easy to handle than the limpid alcoholic lemon extract. Once the user learns how to handle and mix the lemon emulsion into his baking batter, he gets equally good results. Educating the user, however, is a slow, expensive process.

A classical example of consumer education in a new way of using a new product is that of the nonrubbing (self-polishing) floor waxes. Before these were introduced, the best floor wax polishes were pastes, consisting of carnauba and other waxes in turpentine, naphtha, or different solvent mixtures. The standard method of application was to apply the paste to a soft cloth or dauber and spread it over a section of the floor. This then was rubbed and rubbed (using much "elbow grease") until a high polish resulted.

When the nonrubbing (water) waxes were first introduced, the instructions were to wash the floor, mop and dry it and then apply a thin, even coating of the new wax with a clean soft cloth or mop. This sounds perfectly simple, but those who are interested in this field know how many wrong ways were discovered by housewives and maintenance men for applying this wax. Each of these incorrect methods produced a bad result. Consequently, this type of product was slow in gaining favor. Now that the public has learned how to use it, the results are excellent.

TESTING

This book is not intended to give explicit methods of chemical and physical testing. Such methods are known to most chemists and can be found in the standard books on testing. The following general information may prove useful as a starting point in eliminating unsuitable substitute or alternate materials. Data from manufacturers' literature, technical handbooks, or dictionaries, will usually give some information and should be consulted.

Sensual Inspection. Color, clarity, odor, form, homogeneity, grade, etc., can often be checked quickly with a small sample.

Heating a sample in a test tube gives some indication of changes that may be expected in color, odor, form, taste, density, melting point, boiling point, viscosity, vapor pressure, flexibility, cohesion, stickiness, adhesion, drying, slipperiness, elasticity, composition, solvency, stability, bacterial activity, surface tension, explosiveness, flammability, etc.

Strong cooling with dry ice or other freezing mixtures shows changes in many other properties. Manipulation with the fingers gives a quick rough estimate of hardness, flexibility, tenacity or coherence, stickiness, adhesion, feel, slipperiness, elasticity, etc.

A rough estimate of solubility is made by dropping a little of the substitute material and the other ingredients in a test tube with water or with one of the solvents to be used, shaking to see if it dissolves. Solvency is determined in the same way; only in this instance, the solvent is the substitute.

Interaction with other ingredients is tested by using the substitute in the finished product, letting it stand (or even warming it) and determining any apparent change in the properties mentioned in this chapter.

Acidity or alkalinity (pH) is tested with pH papers or a pH meter.

Accelerated Aging Tests.   Whirling a sample in a centrifuge causes the separation of finely divided particles. This separation might not otherwise appear for some time.

The effect of heat is determined by placing a sample in an open or closed tube on a steam or electric plate for a working day or overnight. Heat speeds up most reactions and a test of this nature is often indicative as to how a material will change on aging.

The effect of sunlight can be checked fairly rapidly by exposure to the actinic rays in such devices as the "Fade-o-Meter" or the "Launderometer."

The effect of moisture or dryness is seen by exposure in a closed vessel containing either a dish of water or a desiccant ("Drierite" or sulfuric acid).

The effect of oxygen is rapidly determined by the oxygen bomb test or treatment with an active oxidizing agent such as hydrogen peroxide.

List of Substitutes and Alternatives

The following list must be used with discretion. As previously explained, a substitute for any material may be excellent in one instance and absolutely worthless in another. For example, salt, which has no chemical relationship to acetic acid, is being used instead of the latter in creaming and separating rubber from latex, but it would be useless in a textile "sour" where acidity is a prerequisite. Therefore, any substitute must be tried and tested before commercial use is attempted. Such tests should be made by a competent worker or consultant to avoid subsequent difficulties.

A substitute or alternative need not necessarily be a substance or composition. It can be a process. An example of this is the removal of a metal plating by an electrolytic deplating process or by grinding it off with an abrasive instead of with an acid or other corrosive material.

Some of the listings given are NOT substitutes, but different members of a certain class of products which may be used as alternatives. This serves a dual purpose. First, it shows the representative commercial materials of one group so that available materials which may have been overlooked can be seen. Second, if each of the products is looked up individually, its substitutes will be found.

This list has been compiled, not only for the needs of today, but for the future. A material that is freely available today may be scarce or unavailable tomorrow. On the other hand, a material that is scarce or unobtainable today may be available tomorrow. Furthermore, what may be scarce in the United States may be easily obtained in another country; carnauba wax in Brazil, for example.

HOW TO USE THIS LIST

After looking up the substitutes for a given material, look up, in turn, each substitute mentioned. For example, in looking up glycerin, some of the substitutes listed for it are ammonium lactate, dextrin, glucose, methyl cellulose, and many others. By looking up ammonium lactate, dextrin, and all of the other substitutes listed under glycerin, the total number of possible substitutes will be covered.

Sources of supply of chemicals and allied products can be obtained from the following publishers:

Chemical Industries	New York, NY
Chemical Marketing Reporter	New York, NY
Chemical Week	New York, NY
Drug & Cosmetic Industry	New York, NY
Metals and Alloys	New York, NY
Chemical Catalog Company	New York, NY
Modern Plastics	New York, NY
Soap, Cosmetics Chemical Specialties	New York, NY
Cosmetics & Toiletries	Wheaton, IL
Industrial Chemical News	New York, NY

Product	Substitute or Alternative
Abietic Acid	Rosin
Abrasives	Aluminum oxide, fused
	Bentonite
	Boron carbide
	Carborundum
	Cerium oxide
	Chalk
	Corn cobs, ground
	Crocus
	Cuttlefish bone
	Diamond, industrial
	Diatomaceous earth
	Emery
	Flint
	Fuller's earth
	Garnet
	Iron oxide, red
	Nut shells, ground
	Pumice
	Quartz
	Rottenstone
	Rouge
	Sand
	Silica
	Silicon carbide
	Tripoli
	Walnut shells, ground
Absorbents	"Avicel"
	"Cellite"
	Clay
	Silica, amorphous
	"Zeolex"
Absorption Bases	Cholesterol with lanolin
	See Emulsifiers

Product	Substitute or Alternative

Product

Substitute or Alternative

Accroides, Gum

See Resins
Rosin
Seed-Lac
"Vinsol"

Acetaldehyde

Formaldehyde
Furfuraldehyde
Glyoxal

Acetamide

Ammonium acetate
Ethanolamine acetate
Formamide
Urea

Acetic Acid

See Acids
Ammonium sulfate with dilute sulfuric acid
Boric acid
Citric acid
Formic acid
Gluconic acid
Glycollic acid
Lactic acid
Levulinic acid
Phosphoric acid
Propionic acid
Pyroligneous acid
Saccharic acid
Salt
Sodium bisulfite
Sodium diacetate
Sulfuric acid, dilute
Tartaric acid
Vinegar

Acetone

Butyl alcohol, tertiary
"Cellosolve" with alcohol
Diacetone

Product	Substitute or Alternative
Acetone *(cont'd.)*	Ethyl acetate with isopropyl acetate Isopropyl ether Methyl acetone Methyl ethyl ketone *See* Solvents
Acetylene	Butane
Acetylene Black	Boneblack Carbon black
Acetylene Tetrachloride	*See* Solvents *See* Tetrachlorethane
Acids	Abietic acid Acetic acid Adipic acid Ascorbic acid Azelaic acid Benzene sulfonic acid Boric acid Butyric acid Capric acid Caprylic acid Chloracetic acid Chromic acid Citric acid Cresylic acid Fatty acid Formic acid Fumaric acid. Gallic acid Gluconic acid Glycollic acid Hydrochloric acid Hydrofluoric acid Lactic acid

Product	*Substitute or Alternative*
Acids *(cont'd.)*	Lignosulfonic acid
	Maleic acid
	Malic acid
	Nitric acid
	Oleic acid
	Oxalic acid
	Phosphoric acid
	Propionic acid
	Pyrogallic acid
	Salicylic acid
	Succinic acid
	Sulfamic acid
	Sulfonates, acid
	Sulfuric acid
	Tannic acid
	Tartaric acid
	Toluene sulfonic acid
	Xylene sulfonic acid
Aconite	Isobutyl *p*-aminobenzoate
Acrylonitrile	*See* Elastomers
	Styrene
Adeps Lanae	Absorption bases
	Degras
	Lanolin alcohols
	Petrolatum
Adhesives	Acrylic emulsions
	Albumen
	Balata resin
	Balsam
	Casein
	Cellulose esters, solvent
	Cyanoacrylates
	Dextrin

Product	Substitute or Alternative
Adhesives *(cont'd.)*	Elastomer, solvent
	Epoxy resins
	Flaxseed mucilage
	Glue
	Gluten
	Gums, water dispersible
	Latex
	Lignin compounds
	Pitch
	Polybutenes
	PVAc emulsion
	Pyroxylin
	See Resins, natural
	See Resins, synthetic
	Rosin oil
	Rubber, solvent
	Sodium lignin sulfonate
	Sodium silicate
	Starch
	Tar, soft
	Tincture of benzoin
Adipic Acid	*See* Acids
Adsorbents	Carbon, active
	Clay
	Infusorial earth
	Molecular sieves
	Silica, fumed
	Silica, hydrogel
Agar	Albumen
	Chondrus
	CMC
	Egg white
	Gelatin
	Gums, water dispersible

Product	*Substitute or Alternative*
Agar *(cont'd.)*	Irish moss
	Isinglass
	"Klucel"
	Methyl cellulose
	Mucin
	"Natrosol"
	Pectin
	Silica gel
	Sodium alginate
	Sodium caseinate
	Sodium cellulose glycollate
	Sodium CMC
	See Thickeners
Agave Fiber	Coir
	Cotton
	Esparto grass
	Fibers, synthetic
	Istle fiber
	Jute
Albumen	*See* Acacia
	See Adhesives
	Agar
	Casein
	CMC
	See Emulsifiers
	Gelatin
	Irish moss
	Isinglass
	"Klucel"
	"Natrosol"
	Protein, fish
	Protein, soybean
	See Resins, natural
	See Resins, synthetic
	Sodium CMC
	See Thickeners

Product	*Substitute or Alternative*
Alcohol	*See* Ethyl alcohol
Algicides	"Amine D"
	"Arquad" B-100
	"Bardac" 205M
	"Barquat" 4280
	"BTC" 1100-R
	Chlorine dioxide
	"Hyamine" 1622
Alkalies	Aluminum hydroxide
	Amines, primary, secondary, tertiary, quaternary
	Aminoalcohols
	Ammonium hydroxide
	Barium hydroxide
	Borax
	Calcium hydroxide
	Calcium oxide
	Hydrazine
	Lithium hydroxide
	Magnesium hydroxide
	Magnesium oxide
	Nephelin
	Potassium carbonate
	Potassium hydroxide
	Potassium silicate
	Sodium carbonate
	Sodium hydroxide
	Sodium metasilicate
	Sodium orthosilicate
	Sodium pyrophosphate
	Sodium silicate
	Trisodium phosphate
Alkyd Resins	*See* Resins
Alloys	*See* Metals

Product	*Substitute or Alternative*
Almond Oil	Apricot kernel oil Cherry kernel oil Mineral oil, refined Peach kernel oil "Persic" oil Vegetable oils
Aloe	*See* Thickeners
Aloes	Sucrose octa-acetate *See* Thickeners
Alum	Alum, potash
Alumina	*See* Abrasives
Alumina, Activated	Carbon, activated
Alumina, Fused	Carborundum Fused quartz
Aluminum Acetate	Aluminum chloride Aluminum formate Aluminum sulfate
Aluminum Boro-Tartrate	Aluminum citrate Zinc sulfocarbolate
Aluminum Citrate	Alum Aluminum acetate Aluminum boro-tartrate
Aluminum Ethylate	Magnesium ethylate Potassium ethylate Sodium ethylate

Product	*Substitute or Alternative*
Aluminum Formate	Aluminum acetate Aluminum chloride Aluminum sulfate
Aluminum Hydrate	Alum Ammonia alum Lime Sodium aluminate
Aluminum Hydroxide	*See* Aluminum hydrate
Aluminum Oleate	Calcium oleate, palmitate, resinate, or stearate Lead oleate, palmitate, resinate, or stearate Magnesium oleate, palmitate, resinate, or stearate Zinc oleate, palmitate, resinate, or stearate
Aluminum Oxide	*See* Abrasives
Aluminum Phosphate	Calcium phosphate
Aluminum Powder	Graphite Mica Pearl essence Sericite Slate powder
Aluminum Resinate	Aluminum oleate Barium stearate Calcium resinate Lead resinate Magnesium resinate Zinc palmitate

Product	*Substitute or Alternative*
Aluminum Stearate	"Acrawax" C
	Aluminum oleate
	Barium stearate
	Calcium resinate
	Calcium stearate
	Magnesium stearate
	Manganese stearate
	Paraffin wax
	Soap
	Talc
	Zinc palmitate
	Zinc stearate
Alunite	Bauxite
Amaranth	Cudbear
Amber	Copal, gum
	See Plastics
	See Resins, synthetic
Amides	Acetamide
	Cyanamide
	Dicyandiamide
	Fatty acid amides
	Stearamide
Amines	*See* Alkali
	Amines quaternary
	Ethylamine
	Ethanol amines
Amino Acids	Yeast
Aminoacetic Acid	*See* Glycine
Ammonia Alum	Alum, potash
	Aluminum hydrate

Product	Substitute or Alternative
Ammonia, Anhydrous	Methyl chloride Nitrogen Urea
Ammonia, Aqua	*See* Ammonium hydroxide
Ammonium Acetate	Acetamide
Ammonium Bicarbonate	Carbon dioxide Sodium bicarbonate
Ammonium Chloride	Manganese chloride Zinc chloride
Ammonium Compounds	*See* Alkalies Amides Amines Ammonium thiocyanate Cyanamide Dicyandiamide Urea
Ammonium Hydroxide	*See* Alkalies *See* Amines
Ammonium Lactate	Ammonium glycollate Glycerin
Ammonium Phosphate	Potassium phosphate Sodium phosphate
Ammonium Sulfamate	Borax with boric acid
Ammonium Sulfate	Ammonium phosphate
Ammonium Sulfite	Potassium bisulfite with ammonia Sodium bisulfite with ammonia

Product	*Substitute or Alternative*
Amyl Acetate	*See* Solvents
Amyl Alcohol	Capryl alcohol
	Fusel oil
	Hexyl alcohol
	Octyl alcohol
	See Solvents
	Tetrahydrofurfuryl Alcohol
Annatto	Dyes
	Vegetable colors
Anticaking Agents	"Aerosil"
	"Avicel"
	"Cab-O-Sil"
	Calcium stearate
	Diatomaceous earth
	Silica, fine
	Silicones
	Starch
	Stearic acid
	Tricalcium phosphate
	"Zeothix"
	Zinc stearate
Antimony	Cadmium
	Calcium
	Selenium
	Silver
	Tellurium
Antimony Lactate	Tartar emetic
Antimony Oxide	Molybdenum oxide
	Tin oxide
	Titanium oxide
	Zirconium oxide

Product	*Substitute or Alternative*
Antioxidants	Ascorbyl palmitate
	Alkylaryl phosphites
	Butyl hydroxy anisole
	"CAO"-5
	"Coviox"
	"Cyanox"
	Dihydrocoumarin
	Ethoxyquin
	"Good-Rite"
	Hindered phenols
	Isonox
	Propyl gallate
	t-Butyl hydroquinone
Antiseptics	*See* Preservatives
Antistats	"Ammonyx"-27
	"Arkanlure" Q
	"Armostat" 310
	"Aston" 25
	"Ceraphyl" 65
	"Cordex" DJ
	"Intrasoft" OCN
	"Miranol" DM
	"Monafax" 781
	"Nopcostat"
	"Polyquart"
	"Varstat" 66
	"Witconol" MST
Apricot Kernel Oil	Almond oil
Argon	Helium
Arrowroot	*See* Starch
	See Thickeners

Product	*Substitute or Alternative*
Asbestos	Calcium silicate
	Ceramic fibers
	Cork
	Ebonite
	See Fillers
	Glass fibers
	Magnesia
	Magnesium silicate, fibrous
	Mica
	Mineral wool
	Peat moss
	Refractories
	Silica aerogel
	Slagwool
	Slate
	Talc
	Vermiculite, expanded
	Vulcanized fiber
	Wood wool
Asphalt	Coal tar
	Petroleum sludge
	Pitch, fatty acid
Babassu Oil	Coconut oil
Babbitt	Camwood
	Indium plated iron
	Iron, powdered pressed
	Lead arsenic alloy with ¾% tin
	Lignum vitae
	Silver lead
Bactericides	*See* Antiseptics
	"Bardac" 205M
	Benzalkonium chloride
	"Biopal" NR

Product	*Substitute or Alternative*
Bactericides *(cont'd.)*	"Bretol"
	"Bromal"
	"Bronopol"
	"BTC"
	"Cetol"
	"Cetramide"
	Cetylpyridinium chloride
	Chlorophenol
	Chloroxylenol
	"Dowicide"
	"Dowicil"
	"G-4"
	"Germall" II
	"Glydant"
	Hexylresorcinol
	"Hyamine"
	"Kathon" CG
	"Liquapar"
	Molecular sieves
	"Myacide"
	"Mytab"
	"Nopcocide"
	"Onyxide"
	"Ottasept"
	Paraben
	Potassium sorbate
	See Preservatives
	Propionic acid
	Quaternium
	"Roccal"
	Sodium dehydroacetate
	Sorbic acid
	Tetradecyltrimethyl ammonium bromide
	Tribromosalan
	Trichlorophenol
	"Vancide" 51
	"Variquat" 50

Product	*Substitute or Alternative*
Balata	Chlororubber Gutta percha *See* Rubbers, synthetic
Balsa Wood	Elastomers, foamed "Foamglas" Sponge rubber, hard
Balsam	*See* Resins *See* Tackifiers
Barium	Calcium Magnesium
Barium Carbonate	Barium sulfate Witherite
Barium Hydroxide	*See* Alkalies Calcium hydroxide
Barium Stearate	Aluminum stearate Calcium resinate Calcium stearate
Barium Sulfate	Barium carbonate
Barium Sulfide	Calcium sulfide
Batching Oil	Petroleum scale wax
Bauxite	Alunite Clay, calcined Kaolin, high alumina
Bayberry Wax	Diethylene glycol distearate Japan wax *See* Wax

Product	*Substitute or Alternative*
Beeswax	Ceresin with soap
	Coffee wax
	Esters, fatty
	Flax wax
	Glycerol and glycol fatty acid esters
	Rice wax
	Sugar cane wax
	Triglycerides
	See Wax, synthetic
Belladonna	Atropine
	Stramonium
Bentonite	Clay, colloidal
	See Emulsifiers
	See Fillers
	See Gums, water dispersible
	See Thickeners
Benzaldehyde	Bitter almond oil
	Nitrobenzol
Benzene	*See* Benzol
Benzene Sulfonic Acid	Phenolsulfonic acid
Benzine	Petroleum ether
Benzoic Acid	Cumic acid
	p-Hydroxybenzoic acid
	See Preservatives
Benzoin	Balsams Peru and tolu with trace of vanillin
	Southern sweet gum
Benzol	Ether, petroleum

Product	*Substitute or Alternative*
Benzol *(cont'd.)*	Gasoline *See* Solvents
Benzoyl Peroxide	Hydrogen peroxide
Benzyl Alcohol	*See* Plasticizers
Benzyl Benzoate	*See* Plasticizers Propylene glycol
Beryllium	Lithium
Beta Naphthol	Pyrogallol Resorcinol
Biacetyl	*See* Diacetyl
Binders	*See* Adhesives
Biocides	*See* Preservatives
Birch Tar Oil	Cade oil
Bitter Almond Oil	Benzaldehyde
Bitumen	*See* Asphalt
Blanc Fixe	*See* Fillers *See* Pigments
Bleaches	Bromine Chlorine dioxide Hydrogen peroxide Potassium binoxalate Sodium bisulfite Sodium bromate Sodium chlorite

Product	*Substitute or Alternative*
Bleaches *(cont'd.)*	Sodium hydrosulfite
	Sodium isocyanurate
	Sodium perborate
	Sulfur dioxide
	Zinc hydrosulfite
Bleach, Optical (Whitening Agent)	"Aclarat"
	"Concoflor"
	"Dergopal"
	"Eastman" OB-1
	"Eccobrite"
	"Hiltamine"
	"Intrawite"
	"Leucophor"
	"Tinopal"
	"Uvitex"
Blood Albumen	*See* Adhesives
	Casein
Blood, Human	Bovine serum albumen
Bone	*See* Plastics
Boneblack	Carbon
	Charcoal
	Fuller's earth
	Pitch
	Tar
Boracic Acid	*See* Boric acid
Borax	*See* Alkalies
	Rasorite
Bordeaux Mixture	*See* Insecticides

Product	*Substitute or Alternative*
Boric Acid	*See* Acids
	Phosphoric acid, dilute
Boron Carbide	*See* Abrasives
Bort	*See* Abrasives
Brazilwood	*See* Hypernic
Brea Gum	Gum acacia
British Gum	*See* Dextrin
Bromeline	Keralin
Bromine	Chlorine
	Iodine
Bulking Agents	*See* Fillers
	See Thickeners
Burlap	Kraft paper laminated with asphalt
Butane	Acetylene
	Ethylene
	Gas, manufactured city
	Gas, natural
	Propane
Butanol	*See* Butyl alcohol, normal
Butter	Glyceryl oleo myristate
	Hydrogenated citrus oils
	Hydrogenated fish oils
	Hydrogenated vegetable oils
	Lard
	Mineral oil
	Petrolatum

Product	Substitute or Alternative
Butyl Acetate	*See* Solvents
Butyl Acetyl Ricinoleate	Diglycol ricinoleate with "Carbitol"
Butyl Alcohol, Normal	Butyl alcohol, tertiary with isopropanol *See* Solvents
Butyl Alcohol, Tertiary	Acetone Isopropyl alcohol *See* Solvents
Butyl "Carbitol"	*See* Solvents Triacetin
Butyl "Cellosolve"	*See* Solvents
Butyl Glycollate	*See* Solvents
Butyl Lactate	Glycol glycollate *See* Solvents
Butyl Oleate	Benzyl alcohol Glycol oleate
Butyl Propionate	*See* Solvents
Butyl Stearate	Benzyl alcohol Diglycol laurate Glyceryl monoricinoleate with a little stearic acid
Butylene Glycol	Glycerin Glycol
Butyric Acid	Acetic acid Propionic acid

Product	Substitute or Alternative
Cacao Butter	Borneo tallow Cetyl alcohol with mineral or vegetable oil Hydrogenated oil with beeswax Hydrogenated vegetable oils, partially Monoglycerides with vegetable oil Propylene glycol stearate
Cade Oil	Birch tar oil
Calcium Arsenate	*See* Insecticides
Calcium Carbonate	Calcium phosphate Chalk Clay Dolomite *See* Fillers Gypsum Limestone Marble dust
Calcium Chloride	Alumina, activated Barium monoxide, porous Calcium sulfate, anhydrous Copper sulfate, anhydrous "Florite" Lithium chloride Magnesium chlorate Magnesium chloride Potassium acetate Silica gel Sodium hydroxide, anhydrous
Calcium Flouride	*See* Fluorspar
Calcium Gluconate	Calcium levulinate

Product	Substitute or Alternative
Calcium Hypochlorite	*See* Bleaches
Calcium Levulinate	Calcium gluconate
Calcium Oleate	Aluminum oleate Calcium palmitate Magnesium resinate
Calcium Oxide	*See* Alkalies *See* Lime
Calcium Peroxide	Hydrogen peroxide Sodium peroxide
Calcium Phosphate	Aluminum phosphate
Calcium Propionate	*See* Preservatives
Calcium Resinate	Aluminum resinate
Calcium Stearate	Aluminum stearate
Calcium Sulfate	*See* Gypsum
Camphor	Benzyl benzoate Camphene Dibenzyl Dibutyl tartrate Hexachloroethane Menthol Naphthalene Paradichlorbenzene
Camphor Oil	Turpentine
Candelilla Wax	*See* Wax
Cane Sugar	*See* Sugar

Product	Substitute or Alternative
Capric Acid	Coconut oil fatty acids
Capryl Alcohol	*See* Octyl alcohol, normal
Caramel Coloring	Dyes Malt extract Molasses
"Carbitol"	Monoglycollin *See* Solvents Triacetin
Carbolic Acid	*See* Phenol
Carbon, Activated	Alumina, activated Bone char Carbon, cherry-pit Carbon, coconut shell Carbon, peach-pit Carbon, walnut shell Carbon, wood sawdust Cellulose, fine powder Clay Infusorial earth Magnesium carbonate Magnesium silicate Silica gel Sodium alumino-silicate Talc Wood char
Carbon Bisulfide	*See* Solvents
Carbon Black	Asphalt Boneblack Bone char Charcoal

Product	Substitute or Alternative
Carbon Black *(cont'd.)*	Iron oxide black Lampblack Mineral black Mineral rubber Pitch
Carbon Disulfide	*See* Solvents
Carbon Tetrachloride	Chloroform Ether, petroleum Ethylene dichloride with sulfur dioxide Methyl bromide Methyl chloride *See* Solvents Trichlorethylene
Carborundum	*See* Abrasives
"Carbowax"	Polymerized glycol stearate *See* Wax
Carnauba Wax	"Acrawax" Candelilla wax Cotton wax, green Esparto wax Hydrogenated castor oil Ouricuri wax Stearamides, substituted Sugar cane wax *See* Wax
Carob Gum	*See* Gums, water dispersible *See* Thickeners
Carragheen	*See* Gums, water dispersible *See* Thickeners

Product	Substitute or Alternative

Casein

See Adhesives
Albumen
Gluten
Gums, water dispersible
See Resins, natural
See Resins, synthetic
Shellac
See Thickeners
Whey, dried
Zein

Castor Oil

Diglycol laurate
Glyceryl monoricinoleate
Glycol hexaricinoleate
Grapeseed oil
Vegetable oils

Castor Oil, Dehydrated

Linseed oil, activated

Catalysts

Aluminum ethylate
Boron fluoride
Carbon, activated
Caustic soda
Chlorine
p-Cymene sulfonic acid
Ferric chloride
Hydrochloric acid
Hydrofluoric acid
Iodine
Palladium
Platinum black
Platinum gauze
Rhodium
Sulfuric acid
p-Toluene sulfonic acid
Vanadium pentoxide
Zinc chloride

Product	Substitute or Alternative
Catechin	Mahogany sawdust
Catechu	Mahogany sawdust
Caustic Potash	*See* Alkalies
Caustic Soda	*See* Alkalies
Cedarwood Oil	Naphthalene
Cellophane	Cellulose acetate Cellulose esters Films, plastic Films, synthetic resin Parchment paper *See* Plastics Varnished paper Waxed paper
"Cellosolve"	*See* Solvents
"Cellosolve" Acetate	*See* Solvents
Celluloid	*See* Plastics
Cellulose	Bagasse Cotton linters Fibers, synthetic floss *See* Fillers Paper pulp Rice hulls Sawdust Straw Tanbark, spent Wood flour
Cellulose Esters	*See* Adhesives

Product	Substitute or Alternative
Cellulose Esters *(cont'd.)*	Casein
	See Plastics
	See Resins, natural
	See Resins, synthetic
	Starch acetate
	Sucrose acetate
	Viscose
Ceramic Insulators	Alumina
	Glass
	See Plastics
	Stone
Ceresin Wax	*See* Wax
Cetyl Alcohol	Lanolin alcohols
	Monostearin
	Oleyl alcohol
	Stearyl alcohol and mineral oil
	See Wax
Chalk	*See* Abrasives
	See Calcium carbonate
	See Fillers
Charcoal	Boneblack
	Carbon, activated
	Coal
	Coke
	Peat, dried
	Pitch
	Tar
Chelating Agents	*See* Sequestering Agents
Cherry Gum	*See* Gums, water dispersible
Cherry Kernel Oil	Almond oil

Product	Substitute or Alternative
Chicle	*See* Elastomers *See* Resins
China Clay	*See* Fillers Talc
China Wood Oil	*See* Tung oil
Chinese Blue	*See* Iron blue
Chinese Wax	*See* Insect wax
Chloramin-T	*See* Preservatives
Chorine	Bleaching powder Bromine Chlorine dioxide Hydrogen peroxide Iodine Nitric acid Ozone Sodium chlorite Sodium dichlorisocyanurate Sulfur dioxide Thiourea oxide
Chlorobenzene	*See* Solvents
Chloroform	Carbon tetrachloride *See* Solvents
Chlorophyll	Dyes
Chloropicrin	Furoylchloride *See* Insecticides Methyl bromide
Chlororubbers	*See* Rubber, synthetic

Product	Substitute or Alternative
Chlorthymol	*See* Preservatives
Chlorxylenol	*See* Preservatives
Cholesterol	Lanolin alcohols
Chondrus	Agar *See* Thickeners
Chromates	Molybdates
Chrome Alum	Alum, potash
Chrome Orange	Ochre Orange mineral *See* Pigments Sienna
Chromic Acid	Nitric acid
Chromic Anhydride	*See* Chromic acid
Chromite	Zirconium silicate
Chromium Acetate	Aluminum sulfate
Chromium Potassium Sulfate	*See* Chrome alum
Chromium Sulfate	Ferric sulfate
Chromium Trioxide	*See* Chromic acid
Cinnamon	Cinnamaldehyde and eugenol with powdered nut shells
Citral	Lemongrass oil

Product	*Substitute or Alternative*
Citric Acid	*See* Acids
	Acetic acid
	Gluconic acid
	Glycollic acid
	Lactic acid
	Lemon juice
	Levulinic acid
	Malic acid
	Phosphoric acid
	Propionic acid
	Saccharic acid
	Sodium acid sulfate
	Sodium bisulfite
	Sodium diacetate
	Tartaric acid
	Sulfuric acid, dilute
	Vinegar
Citronella Oil	Eucalyptus oil
	Pennyroyal oil
	Tetra hydrofurfuryl lactate
Clay	*See* Fillers
	See Thickeners
Clay, Colloidal	*See* Bentonite
	Kaolin
Clove Oil	Eugenol
Cobalt Naphthenate	*See* Driers
Cobalt Oxide	Sodium antimonate, manganate, uranate, or vanadate with copper carbonate
Cochineal	Dyes, aniline
	Vegetable colors

Product	*Substitute or Alternative*
Cocoa	Carob beans, flavor
Cocoa Butter	*See* Cacao butter

Coconut Oil	Babassu oil
	Castor oil
	Castor with cottonseed oils
	Coconut oil fatty acids
	Cohune oil fatty acids
	Corozo oil
	Glyceryl fatty acid esters
	Glyceryl myristate with castor oil
	Hydrogenated vegetable oils, partially
	Lard oil
	Mineral oil with lard oil
	Myristic with ricinoleic acids
	Neatsfoot oil
	Oleic with ricinoleic acids
	Olive oil with mineral oil
	Oxidized paraffin wax with red oil
	Palm kernel oil
	Peanut oil
	Polyhydric alcohol, fatty acid esters, e.g., diglycol ricinoleate
	Rosin with linseed oil
	See Vegetable oils

Cod Liver Oil	Rice bran oil, purified
	Sardine oil
	Shark liver oil
	Tuna liver oil

Cod Oil	Degras
	Hake liver oil
	Herring oil, blown
	Menhaden oil, blown
	Pilchard oil

Product	*Substitute or Alternative*
Cod Oil *(cont'd.)*	Sardine oil, blown
	Whale oil
Coffee	Bicho seeds
	Chickory
	Grains, mixed roasted
	Nuts, roasted
Coffee Wax	Beeswax
	See Wax
Colchicine	Sanguinarin
Collodion	*See* Adhesives
	Cellulose esters solutions
	Plastics solutions
	Resins, solution of synthetic
Colophony	*See* Rosin
Colors	*See* Dyes
	See Pigments
Congo, Gum	*See* Copal
Copal	*See* Adhesives
	See Resins, natural
	See Resins, synthetic
Copper Chromate	Creosote
	See Preservatives
Copper Naphthenate	Copper carbonate, basic
	Copper "mahogany" sulfonates
	Copper oleate
	Creosote
	Driers
	See Preservatives
	Tar oil

Product	*Substitute or Alternative*
Copper Oxide	Manganese dioxide *See* Pigments
Copper Sulfate	*See* Preservatives
Copper Tungsten	Silver molybdenum carbide
Cork	Asbestos fiber with asphalt or resin binder Bark fiber with asphalt or resin binder Bran fiber with asphalt or resin binder Felt, impregnated hair or wool Foamed elastomers Foamed glass Frothed and set synthetic resins Linseed meal Millboard, soft Mineral wool Oatmeal Palmetto wood Paper pulp Peat moss Redwood bark Rock "wool" Rubber, foamed *See* Rubbers, synthetic Rubber with lignin Sawdust Silica aerogel Sphagnum moss with asphalt or resin binder Wood wool
Corn Sugar	*See* Sugar
Corn Syrup	Glycerin *See* Sugar

Product	Substitute or Alternative
Corrosion Inhibitors	"Actramide"
	"Alox"
	Anthranilic acid
	"Arcor"
	"Arnox"
	"Arquest"
	"Arzoline"
	"Butoxyne"
	Butynediol
	"Cachalot"
	"Cobratec"
	"Cyclophos"
	"Dequest"
	"Drewgard"
	"Emulsogen"
	"Hostacor"
	"Kemamine"
	"Miramine"
	"Miranol"
	"Molykote"
	"Moly White"
	"Monacor"
	"Monalube"
	"Monamulse"
	"Morlex"
	Morpholine
	"Nocap"
	"Nopcogen"
	"Petrobase"
	Petrolatum
	"Polyrad"
	Propargyl alcohol
	"Rotax"
	"Solar"
	Tannic acid
	"Textamine"
	"Tomah" PA

Product	Substitute or Alternative
Corrosion Inhibitors *(cont'd.)*	"Unamine" O "Witcamine" RAD "Witcor"
Corundum	*See* Abrasives
Cotton	Cellulose Fibers, synthetic
Cotton Linters	Alpha cellulose
Cotton, Soluble	*See* Nitrocellulose
Coumarin	Melilotin Vanillin
Cream of Tartar	Adipic acid Ammonium sulfate Mucic acid Saccharolactic acid
Creosote	Coal tar Copper chromate Copper naphthenate Copper oleate Copper phosphate Copper sulfate Cresylic acid Pentachlorphenol *See* Preservatives Tar oils Zinc chloride
Cresol	Coal tar acids Creosote Furfural Phenol *See* Preservatives

Product	*Substitute or Alternative*
Cresylic Acid	Creosote *See* Cresol
Crocus	*See* Abrasives Iron oxide
Cryolite	Cryolite, synthetic Sodium silicofluoride
Cube Root	*See* Insecticides
Cudbear	Amaranth Dyes
Cumic Acid	Benzoic acid
Cutch	*See* Dyes Logwood Mahogany sawdust
Cuttlefish Bone	*See* Abrasives
Cyclohexanol	*See* Solvents
Cyclohexanone	*See* Solvents
Cymene	*See* Solvents
Damar	*See* Resins
Decolorizing Carbon	*See* Carbon, activated
Defoamers	"Acrawax" C dispersion "Actrafoam" "Actrasol" "Albon" "Arsil"

Product	Substitute or Alternative

Defoamers *(cont'd.)*

"Balab" Bubble Breaker
"Chemax"
"Depuma"
"Dow-Corning" Silicone 1500
"Exfo"
"Fancol"
"Fleetcol"
"Foamaster"
"Foamex" A
"Foamgard"
"Lipowax" C
"Mazu" DF
Mineral oil
"Nilofoam"
"Pamelyn"
"Patcote"
"Perenol"
"Poly-G Fluid"
"Rexfoam"
"Ridafoam"
"Sag"
Silicones
Simethicone
"Surfynol" 104
"Surpasol"
"SWS"

Degras

Cod oil
Petrolatum and rosin oil
See Wool grease

Deodorants

Chlorine dioxide
Thiourea oxide

Derris

Devil's shoestring root
See Insecticides

Product	*Substitute or Alternative*
Dessicants	Alumina, activated
	Calcium chloride
	Silica gel
	Starch acrylate
Detergents	*See* Soaps
	See Wetting agents
Dextrin	*See* Adhesives
	See Gums, water dispersible
	Malt extract
	Molasses
	See Resins
	Sodium silicate
	Sugar
	See Thickeners
	Urea-formaldehyde
Dextrose	*See* Sugar
Diacetone	Acetone
	See Solvents
Diacetone Alcohol	*See* Solvents
Diacetyl	Acetyl methyl carbinol
Diamond, Industrial	*See* Abrasives
	Boron carbide
	Carborundum
	Sapphire, synthetic
	Silicon carbide
Diamyl Phthalate	*See* Plasticizers
Diatomaceous Earth	*See* Abrasives
	See Fillers
	Fuller's earth

Product	*Substitute or Alternative*
Dibenzyl	Camphor
Dibutyl Phthalate	Butyl oleate Castor oil Castor oil, blown Glycol hexaricinoleate 2,5 Hexanediol Monoglycollin *See* Plasticizers
Dichloramine	Chlorine dioxide "Dantoin" Hydrogen peroxide
Dichlorethylene	Methyl chloride *See* Solvents
Dichlorodiethyl Ether	*See* Solvents
Dicresyl Carbonate	Butyl oleate Dibutyl phthalate Glycerin *See* Plasticizers
Diethyl "Carbitol"	*See* Solvents
Diethyl Carbonate	*See* Solvents
Diethylene Glycol	*See* Glycols
Diglycol Phthalate	*See* Plasticizers
Diglycol Stearate	*See* Emulsifiers Glycol esters Gums, water dispersible *See* Soap Wax, emulsified

Product	*Substitute or Alternative*
Dioxan	*See* Solvents
Diphenyl Oxide	*See* Plasticizers
Disinfectants	*See* Bactericides
Dispersing Agents	Sodium lignin sulfonate *See* Surfactants
Divi-Divi	Cascolate pods Tara
Dolomite	Calcium carbonate Clay Limestone Magnesite
"Dowicides"	*See* Preservatives
Dragon's Blood	*See* Dyes
Driers	Cobalt linoleate Cobalt naphthenate Cobalt resinate Lead linoleate Lead naphthenate Lead resinate Magnesium oleate Manganese linoleate Manganese naphthenate Manganese resinate Zinc palmitate
Dyes	"A.A.P. Naphthols" Alkanet root Amaranth Annatto

Product	*Substitute or Alternative*
Dyes *(cont'd.)*	Caramel coloring
	Chlorophyll
	Cochineal
	Coffee grounds
	Cudbear
	Cutch
	Dragon's blood
	Dyes, aniline
	Fustic
	Hypernic
	Indigo
	Lignin sulfonates
	Logwood
	Madder
	Orchil extract
	Osage orange extract
	Pigments, mineral, e.g., sienna
	Pitch
	Precipitates, chemical, e.g., antimony sulfide
	Precipitate, chemical, e.g., lead chromate
	Quercitron bark
	Saffron
	Tannin
	Tar oil
	Tea leaves
	Turmeric
	Vegetable colors
Ebony	Asphalt compositions
	See Plastics
Egg White	Agar
	Lecithin
	Pectin
	Soyabean protein

Product	*Substitute or Alternative*
Egg Yolk	*See* Emulsifiers Fish milt Lecithin Polyhydrical alcohol fatty acid esters, e.g., glyceryl monostearate
Elastomers	*See* Plastics *See* Rubber, synthetic
Elaterite	*See* Mineral rubber
Emery	*See* Abrasives Aluminum oxide, artificial
Emulsifiers	Albumen Amine soaps, e.g., trihydroxyethyl-amine oleate Bentonite Casein CMC Gelatin Gum, water dispersible Lanolin Lecithin Lignin sulfonates Methyl cellulose Monoglycerides Petroleum sulfonates Polyglycerol fatty esters Polyhydroxyalcohol fatty acid esters, e.g., glyceryl monoricinoleate Saponin Soap Sodium stearoyl lactylate Sorbitol fatty esters Sulfated vegetable oils *See* Thickeners *See* Wetting agents

Product	Substitute or Alternative
Ergot	Huitlacoche (Mexican corn fungus)
Esparto Wax	*See* Wax
Essential Oils	Synthetic aromatics Synthetic oils
Ester Gum	*See* Resins, natural *See* Resins, synthetic Rosin Shellac Soya protein Zein
Ethanolamines	Alkalies Amines
Ether, Ethyl	Ether, petroleum Ethyl chloride Isopropyl ether Methylal Methyl-*t*-butyl ether *See* Solvents
Ether, Petroleum	Benzol Carbon tetrachloride Ether, ethyl Ethyl chloride Isopropyl ether Pentane *See* Solvents
Ethyl Acetate	Acetone Isopropyl acetate Methyl acetate *See* Solvents
Ethyl Alcohol	Glycols

Product	Substitute or Alternative
Ethyl Alcohol *(cont'd.)*	Isopropanol
	Methyl alcohol
	Pentane
	Propylene glycol
	Rum
	See Solvents
	Sulfated oils
	Tetrahydrofurfuryl alcohol
	Wine
Ethyl Butyrate	*See* Solvents
Ethyl Chloride	Ether, ethyl
	Ether, petroleum
Ethyl Ether	*See* Solvents
Ethyl Lactate	*See* Solvents
Ethyl Propionate	*See* Solvents
Ethylene	Butane
	Propylene
Ethylene Chlorhydrin	Glycerylchlorhydrin
Ethylenediamine	*See* Alkalies
	Ammonia
	Morpholine
Ethylene Dichloride	Methyl chloride
	Propylene dichloride
	See Solvents
Ethylene Glycol	*See* Glycols
Excelsior	Coir

Product	*Substitute or Alternative*
Excelsior *(cont'd.)*	Grass, dried
	Paper, shredded
	Peat moss
	Perlite, expanded
	Plastics, foamed
	Sawdust
	Spanish moss
	Straw
	Tanbark, spent
Fatty Acids	Heptanoic acid
	Oxidized paraffin wax
	Pelargonic acid
	Ricinoleic acid
	Rosin oil
	Tall oil
Ferric Chloride	Alum
	Aluminum chloride
	Ferric sulfate
	Magnesium chloride
Ferric Sulfate	Alum, potash
	Ferric chloride
Ferromanganese	Spiegeleisen
Ferrous Chloride	Aluminum chloride
	Barium chloride
Ferrous Sulfate	*See* Copperas
Fibers	Agave
	Asbestos
	Balsam wool
	Bristles
	Cellulose esters
	Coir

Product	*Substitute or Alternative*
Fibers *(cont'd.)*	Cotton
	Esparto grass
	Fique
	Flax
	Glass fibers
	Hair, animal
	Hair, human
	Hemp
	Ixtly fiber
	Mohair
	Nylon
	Paper, braided
	Piassava
	Polyester
	Protein, hardened
	Ramie
	Rayon
	See Resins, synthetic
	Sansevieria
	Silk
	Sisal
	Straw
	Viscose
	Wool
	Yucca
Fillers	Alumina, hydrated
	Asbestos
	Bagasse
	Barytes
	Bentonite
	Bran
	Calcium carbonate
	Calcium sulfate
	Chalk
	Charcoal
	Clay
	Coal powder

Product	*Substitute or Alternative*
Fillers *(cont'd.)*	Cork
	Cotton flock
	Diatomaceous earth
	Dolomite
	Kaolin
	Limestone
	Magnesia
	Magnesite
	Magnesium silicate
	Minerals
	Paper pulp
	Perlite
	Pyrophyllite
	Rayon flock
	Roots, crushed
	Sand
	Sawdust
	Silica
	Slate flour
	Stone, crushed
	Talc
	Whiting
	Wood flour
Fire Clay	Agalmatolite
	Chalcedony
	Mollite
	Pinite
Fish Oils	Animal oils
	Cod liver oil fatty acids
	Herring oil
	Menhaden oil
	Pilchard oil
	Sardine oil
	Vegetable oils
	Whale oil

Product	Substitute or Alternative
Flameproofing	Ammonium salts
	Antimony oxide
	Brominated diphenyl oxide
	Brominated hydrocarbons
	Brominated phenols and glycols
	Chlorinated hydrocarbon
	Chlorinated paraffin
	Organic phosphorated compounds
	Phosphates, inorganic
Flavoring Extracts	Imitation flavors
Flax	Cotton with jute
	See Fibers
	Flax tow
Flax Wax	*See* Wax
Fleaseed	*See* Psyllium seed
Flint	*See* Abrasives
Fluorine	Chlorine
	Iodine
Fluorspar	Ammonium bifluoride
	Cryolite
	Sodium silicofluoride
Foambreakers	*See* Defoamers
Formaldehyde	Acetaldehyde
	Furfural
	Glyoxal
	See Preservatives
Formalin	*See* Formaldehyde

Product	*Substitute or Alternative*
Formamide	Acetamide
	Urea
Formic Acid	*See* Acids
"Freon"	Ammonia (anhydrous)
	Carbon dioxide
	Low boiling hydrocarbons
	Methyl chloride
	Methylene dichloride
	See Solvents
	Sulfur dioxide
Fuller's Earth	*See* Abrasives
	Carbon, activated
	Diatomaceous earth
	See Fillers
Fumaric Acid	*See* Acids
	Adipic acid
Fungicide	Lauric acid
	Mercury compounds
	Thiram
	Undecylenic acid
	Zinc undecylenate
Furfural	Acetaldehyde
	Benzaldehyde
	Formaldehyde
Furfuryl Alcohol	*See* Solvents
Furoyl Chloride	Chloropicrin
Fusel Oil	Amyl alcohol
	Butyl and amyl alcohols

Product	Substitute or Alternative
Fusel Oil *(cont'd.)*	*See* Solvents Tetrahydrofurfuryl alcohol
Fustic	*See* Dyes
Garnet	*See* Abrasives
Gas, Natural	Acetylene Butane Ethylene Producer gas Propane
Gas Oil	Butane Gasoline, crude Kerosene
Gasoline	Acetylene with anhydrous ammonia Alcohol Benzol Butane Gasohol Hydrogen Methane Methanol Petroleum ether Producer gas Propane Synthetic gas
Gelatin	Acrylic dispersion *See* Adhesives Agar Alginates Casein CMC Dextrin

Product	*Substitute or Alternative*

Gelatin *(cont'd.)*

See Emulsifiers
Gluten
See Gums, water dispersible
Isinglass
Pectin
Polyvinyl alcohol
See Resins, synthetic
Shellac, alkali
Soaps
Sodium alginate
See Thickeners
Zein

Germicides

See Bactericides

Gilsonite

Manjak
See Mineral rubber
Pitch

Gluconic Acid

See Acids
Citric acid
Lactic acid
Mucic acid

Glucose

See Sugar

Glue

Acrylate emulsions
See Adhesives
Casein
See Emulsifiers
Gelatin
Gums, water dispersible
Latex
Polyvinyl acetate emulsion
Resins, synthetic
Rosin soaps
See Thickeners

Product	*Substitute or Alternative*
Gluten	Casein
	Soy protein
	See Thickeners
Glycerin	"Ajidew"
	Alginates
	Aminoalcohols
	Ammonium lactate
	Apple syurp
	Butylene glycol
	Calcium chloride
	Corn syrup
	Dextrin
	Dibutyl phthalate
	Dicresyl carbonate
	Diglycol oleate
	Ethylammonium phosphate
	Glucose
	Glycols
	Invert sugar
	Lactic acid
	Magnesium chloride
	Methyl cellulose
	Methyl sodium potassium phosphate
	Mineral oil
	Nonaethylene glycol ricinoleate
	Propylene glycol
	Sorbitol syrup
	Sugar
	Sulfated castor oil
Glyceryl Chlorhydrin	Ethylene chlorhydrin
Glyceryl Phthalate	Glyceryl maleate
Glycol Diacetate	Monoglycollin
	See Solvents
	Triacetin

Product	*Substitute or Alternative*
Glycol Glycollate	Butyl lactate
Glycollic Acid	Acetic acid *See* Acid Citric acid Lactic acid
Glycol Monoacetate	Monoglycollin
Glycols	Glycerin *See* Solvents Sorbitol
Glyoxal	Formaldehyde
Grapeseed Oil	*See* Vegetable oils
Graphite	Boneblack with talc Metals, powdered Mica, powdered Molybdenum disulfide Paraffin wax Red lead Silica black Talc *See* Wax
Gums	*See* Adhesives *See* Emulsifiers *See* Resins, natural *See* Resins, synthetic *See* Thickeners
Gum Benzoin	*See* Benzoin
Gum Karaya	*See* Gums, water dispersible
Gum, Locust Bean	*See* Gums, water dispersible

Product	*Substitute or Alternative*
Gum Tragacanth	*See* Gums, water dispersible
	See Thickeners
Gums, Water Dispersible	Acrylic dispersion
	See Adhesives
	Agar
	Alginates
	Carragheen
	Casein
	Cherry gum
	CMC
	Dextrin
	Diglycol stearate
	See Emulsifiers
	Ghatti
	Gluten
	Glutrin
	Guar
	Gum acacia
	Gum karaya
	Gum tragacanth
	Hydroxypropyl methyl cellulose
	Isinglass
	"Klucel"
	Locust bean gum
	Methyl cellulose
	"Natracel"
	Pectin
	"Polyox"
	Psyllium seed
	Quince seed
	Soap
	Sodium alginate
	Sodium borophosphate
	Starch
	See Thickeners
	Whey, dried
	Xanthan gum

Product	*Substitute or Alternative*
Gutta Percha	Balata
	See Elastomers
	See Resins, synthetic
	See Rubber, synthetic
Gypsum	*See* Fillers
Hair	Bristles
	"Dacron"
	Fiberglass
	See Fibers
	Rubber sponge
	Wire, steel
Helium	Argon
	Hydrogen
Hematine Extract	*See* Logwood
Hemp	*See* Fibers
Hempseed Oil	Sunflower oil
	See Vegetable oils
Henna	*See* Dyes
Heparin	3, 3' Methylene-*bis* (4-hydroxycoumarin)
Herring Oil	Cod oil
Hexamethylenetetramine	Hydrofuramide
	Urea
Hexamine	*See* Hexamethylenetetramine
Hexyl Acetate	*See* Solvents

Product	Substitute or Alternative
Hexyl Alcohol	Octyl alcohol, normal
Hormones, Sex	Cafesterol
Horn	See Plastics
Horse Hair	Bristles See Fibers Ixtle
Humectant	"Ajidew" Calcium chloride "Cellasol" Glycerin Glycols Sorbitol
Hydrobromic Acid	See Acids
Hydrochloric Acid	See Acids Hydrobromic acid Nitre cake Sulfuric acid
Hydrocyanic Acid	See Insecticides
Hydrofluoric Acid	Aluminum chloride, anhydrous Ammonium bifluoride Phosphoric with chromic acids Sodium bifluoride Sodium silico fluoride
Hydrofuramide	Hexamethylenetetramine
Hydrogen	Acetylene Helium

Product

Substitute or Alternative

Hydrogen Peroxide

Benzoyl peroxide
Bleaching powder
Calcium peroxide
Chloramine
Chloramine T
Chlorine
Chlorine dioxide
"Dantoin"
Dichloramine
Magnesium peroxide
Oxalic acid
See Oxidizers
Oxygen
Ozone
Potassium bichromate
Potassium chlorate
Potassium chromate
Potassium perchlorate
Potassium permanganate
See Preservatives
Selenium dioxide
Sodium borohydride
Sodium chlorate
Sodium chlorite
Sodium hydrosulfite
Sodium hypochlorite
Sodium perborate
Sodium perchlorate
Sodium peroxide
Sulfur dioxide
Trichlorcyanuric salts
Zinc hydrosulfite
Zinc peroxide

Hydroxyacetic Acid

See Glycollic acid

Hydroxycitronellal

Cyclamen aldehyde

Product	Substitute or Alternative
Hypernic	*See* Dyes
Iceland Moss	*See* Emulsifiers *See* Gums, water dispersible *See* Thickeners
India Gum	*See* Gums, water dispersible
Indigo	*See* Dyes
Infusorial Earth	*See* Fillers
Insecticides	Amides, higher fatty, e.g., lauryl amide Barium silicofluoride Bordeaux mixture Calcium arsenate Castor leaf extract Chloropicrin Cryolite DDT Hydrocyanic acid Ketones, higher fatty Lead arsenate Lindane Methyl bromide Naphthalene Nicotine sulfate Paradichlorobenzene Paris green Phenothiazine Phthalonitrile Pyrethrins Pyrethrum Rotenone Sodium fluoride Sodium silicofluoride

Product	*Substitute or Alternative*
Insecticides *(cont'd.)*	Sulfur
	Tetrahydrofurfuryl lactate
	Tobacco dust
Insect Wax	*See* Wax
Invert Sugar	*See* Sugar
Iodine	Bromine
	Chlorine
	Fluorine
	See Preservatives
Iridium	Ruthenium
	Ruthenium with platinum
Irish Moss	*See* Gums, water dispersible
	See Thickeners
Iron Blue	*See* Pigments, inorganic
Iron Oxide	*See* Abrasives
	Alumina
	Graphite
	See Pigments
	Spanish oxide
Isinglass	Agar
	Alum, potash
	See Emulsifiers
	See Gums, water dispersible
	See Thickeners
Isobutyl *p*-Aminobenzoate	Aconite
Isophorone	*See* Solvents

Product	*Substitute or Alternative*
Isopropyl Acetate	Ethyl acetate *See* Solvents
Isopropyl Alcohol	Butyl alcohol, tertiary *See* Solvents
Isopropyl Ether	Acetone Ether, ethyl Ether, petroleum *See* Solvents
Ivory	Aminoplasts Ivory nut *See* Plastics *See* Resins, synthetic
Ivory, Vegetable	Melamine resins
Japan Wax	*See* Wax
Jet	*See* Plastics
Jute	*See* Fibers Kraft paper, twisted with cotton braid Malva fiber Yucca fiber
Kaolin	Bentonite *See* China clay *See* Fillers
Kapok	Elastomers, foamed Expanded perlite Fiber floss, synthetic Glass foam
Kauri	*See* Resins

Product	Substitute or Alternative
Kieselguhr	*See* Fillers
Kyanite	Alumina, fused Chrome magnesia Kaolin Magnesia alumina (Spinel) Mullite Silica brick Stillimanite
Lactic Acid	*See* Acid Ammonium sulfate with dilute sulfuric acid
Lactose	Dextrose Fructose Milk powder, skimmed *See* Sugar
Lampblack	Boneblack Carbon black *See* Dyes
Lanolin	Petrolatum with rosin Soap, soft
Lanolin Alcohols	Cetyl alcohol Cholesterol Monostearin Olein Phytosterols
Lard	Hydrogenated vegetable or fish oils Tallow, refined
Lard Oil	Fish oil Mineral oil

Product	Substitute or Alternative
Lard Oil *(cont'd.)*	Polyhydric alcohol fatty acid esters, e.g., diglycololeate
	Rosin oil
	See Vegetable oils
Latex	Acrylic emulsions
	See Adhesives
	Elastomer emulsions
	Reclaimed rubber emulsions
	See Resin emulsions
	Rubber emulsions, synthetic
	Vinyl copolymer emulsions
Lauric Acid	*See* Fatty acids
	Stearic acid with castor oil
Lead Arsenate	*See* Insecticides
	Nicotine sulfate with bentonite
	Phthalonitrile
Lead Azide	Lead fulminate
	Mannitol hexanitrite
	Mercury fulminate
Lead Chloride	Barium chloride
Lead Chromate	Yellow iron oxide with a little zinc chromate
	Zinc tetroxy chromate
Lead Fulminate	Lead azide
Lead Linoleate	*See* Driers
Lead Naphthenate	*See* Driers
Lead Oleate	Aluminum oleate

Product	*Substitute or Alternative*
Lead Oleate *(cont'd.)*	Calcium oleate *See* Driers Lead naphthenate
Lead, Red	Graphite Iron oxide, red Iron phosphate *See* Pigments White lead
Lead Wool	Asbestos wool Fibers, resin impregnated Mineral wool
Lead Resinate	Aluminum resinate *See* Driers
Leather	Barks, tree Felt Fiberboard Fibers woven, impregnated, coated or compressed Linoleum Paper, impregnated *See* Plastics Plywood, soft Rubber *See* Rubbers, synthetic 'Vulcanized fiber Wood, plasticized
Lecithin	Betaine di-oleylglycerophosphate Egg yolk *See* Emulsifiers Phosphatides, vegetable oil
Lemon Oil	Citral with limonene or lemon oil terpenes

Product	Substitute or Alternative
Lemongrass Oil	Dipentene Lemon oil terpenes
Levulinic Acid	Acetic acid *See* Acid Citric acid
Levulose	*See* Sugar
Licorice Root	Protein, hydrolyzed *See* Wetting agents
Lignin Sulfonates	*See* Emulsifiers *See* Thickeners
Lime	*See* Alkalies Alum, potash Barium oxide Calcium carbide residue
Limonene	Dipentene Lemon oil terpenes
Linseed Oil	Alkyd resin solutions Fish oils *See* Vegetable oils
Litharge	Red lead
Lithium	Beryllium Calcium
Lithium Chloride	Calcium chloride
Lithium Hydride	Calcium hydride Potassium hydride Sodium hydride

Product	Substitute or Alternative
Lithium Hydroxide	*See* Alkalies
Logwood	*See* Dyes
Lubricants	"Alfol"
	"Alox"
	Calcium stearate
	Castor oil
	Diglycol ricinoleate
	Ethylene *bis*-stearamide
	Fatty acids
	Fatty alkanolamide
	Glycerin
	Glyceryl fatty acid esters
	Glycol fatty acid esters
	Glycols
	Graphite
	"Lipowax" C
	Mineral oil
	Molybdenum disulfide
	Petrolatum
	Polyglycol fatty esters
	Rape seed oil, blown
	Silicones
	See Softeners
	Tallow
	See Wax
Lycopodium	Buck-grass
	Clubfoot moss
	Milkweed spores, dried and powdered
	Pine grass
	Snake moss
	Walnut shell partings, powdered
Madder	*See* Dyes

Product	*Substitute or Alternative*
Magnesia	*See* Alkalies
	Asbestos
	See Fibers
	See Fillers
	Manganese hydroxide
	Peat
	Refractories
	Zinc oxide
Magnesite	Dolomite
	See Fillers
	Magnesium carbonate
Magnesium Carbonate	*See* Fillers
	Talc
Magnesium Ethylate	Aluminum ethylate
Magnesium Hydroxide	*See* Alkalies
Magnesium Oleate	Aluminum oleate
	See Driers
Magnesium Oxide	*See* Magnesia
Magnesium Peroxide	*See* Oxidizers
Magnesium Resinate	Aluminum resinate
	See Driers
Magnesium Silicate, Fibrous	Asbestos
	Ceramic fibers
	Glass fibers
Magnesium Stearate	Aluminum stearate
	See Anticaking agents

Product	Substitute or Alternative
Maize Oil	*See* Corn oil
Maleic Acid	*See* Acids Adipic acid Lactic acid Malic acid Phthalic anhydride Sebacic acid Succinic acid
Malic Acid	*See* Acids
Malonic Acid	*See* Acids
Manganese Chloride	Ammonium chloride
Manganese Dioxide	*See* Oxidizers
Manganese Hydroxide	Lime, slaked Magnesia
Manganese Naphthenate	*See* Driers
Manganese Stearate	*See* Anticaking agents
Manila Fiber	*See* Fibers
Manila Gum	Cumarone resins *See* Resins
Manjak	Mineral rubber Pine tar
Mannitol	Sorbitol
Menhaden Oil	Cod oil Tallow oil

Product	Substitute or Alternative
Menthol	Camphor with peppermint oil
Mercuric Chloride	Copper oxide *See* Preservatives Silver proteinate
Mercury Fulminate	Diazodinitrophenol Lead azide Lead styphnate-hypophosphite Nitromannite
Mesityl Oxide	*See* Solvents
Metaldehyde	Paraformaldehyde
Methanol	*See* Methyl alcohol *See* Solvents
Methyl Acetanilide	Camphor *See* Plasticizers
Methyl Acetate	Ethyl acetate *See* Solvents
Methyl Acetone	Acetone *See* Solvents
Methylal	Ether, isopropyl Formaldehyde *See* Solvents
Methyl Alcohol	Alcohol, denatured *See* Solvents
Methylamine	*See* Amines Ammonia Morpholine

Product	*Substitute or Alternative*
Methyl Bromide	Carbon tetrachloride Chloropicrin *See* Insecticides *See* Solvents
Methyl "Cellosolve" Acetate	*See* Solvents
Methyl Cellulose	Agar Alginates Casein CMC Emulsifiers Glycerin *See* Gums, water dispersible Latex Polyvinyl alcohol *See* Thickeners
Methyl Chloride	Ammonia, anhydrous Carbon dioxide Dichlorethylene Ethylene dichloride "Freon" Methylene chloride Propane *See* Propellants Sulfur dioxide
Methylene Chloride	*See* Solvents
Methyl Ethyl Ketone	Acetone *See* Solvents
Methylheptine Carbonate	Phenylethylphenyl acetate
Mica	Asbestos Ceramics

Product	Substitute or Alternative
Mica *(cont'd.)*	Glass fibers
	Magnesium silicate, artificial
	See Plastics
	Porcelain, electrical
	Pressboard, thoroughly dried and impregnated with oil
Milk	Soyabean milk
Milk Sugar	*See* Lactose
Milori Blue	*See* Iron blue
Mineral Rubber	Asphalt
	See Elastomers, foamed
	Manjak
	Pitch
Mineral Spirits	*See* Solvents
Moellen Degras	*See* Wool grease
Molasses	Corn syrup, flavor
	Dextrin
	Distillery slop
	Milk whey, concentrated
	Sorghum syrup
	Sulfite liquor
Mold Release	"Acrawax" C
	"Alox" 111
	"Arkolube"
	"Armoslip"
	Calcium stearate
	"Cardipol"
	"Drewfax" 169
	"Emphos" D70

Product	*Substitute or Alternative*
Mold Release *(cont'd.)*	Fats
	"Formasil"
	"Lipowax" C
	See Lubricants
	"Masil" 265
	"Mekon"
	Mineral oil
	"Polycone"
	"Polywax"
	Silicones
	Starch
	Stearic acid
	"SWS"-222
	See Vegetable oils
	See Wax
	"Witco" MRC
Monoacetin	Monoglycollin
Monoglycollin	Glycol diacetate
	Triacetin
Montan Wax	Lignite wax
	See Wax
Montmorillonite	*See* Aluminum silicate
Morpholine	Ammonia
	Ethylenediamine
	Methylamine
	Monoethanolamine
	Triethanolamine
Mother-of-Pearl	Melamine resins
	Mica, TiO_2 coated
	Pearl essence
Mucic Acid	*See* Acids

Product	Substitute or Alternative
Mucin	Agar *See* Thickeners
Muscovite	*See* Mica
Musk	Abelmoschus
Mustard Gas	Benzyl dichloride
Mustard Seed Oil	Lard oil *See* Vegetable oils
Myristic Acid	Coconut oil fatty acids with stearic acid *See* Fatty acid
Naphtha	*See* Solvents
Naphthalene	Camphor Cedarwood oil *See* Insecticides Paradichlorobenzol
Naphthenates, Metallic	*See* Driers
Naphthenic Acid	*See* Fatty acids
Neatsfoot Oil	Degras *See* Fish oils Mineral oil Polyhydric alcohol fatty acid esters, mixed Tallow oil *See* Vegetable oils
Nicotine	*See* Insecticides
Nitric Acid	*See* Acids

Product	Substitute or Alternative

Nitrocellulose

See Adhesives
Cellulose esters
See Plastics
See Resins, synthetic
Vinyl copolymers

Nutgalls

Gall apples
Oak galls (oak apples)
Tannin

Nux Vomica

Strychnine hydrochloride

Octyl Alcohol, Normal

Hexyl alcohol
Tributyl phosphate

Odorants

Ammonium polysulfide
Benzaldehyde
Dipentene
Mercaptans
Methyl salicylate
Nitrobenzol
p-Chlorbenzol
Vanillin

Oleic Acid

See Fatty acids
Long chain acids
Tall oil

Oleyl Alcohol

Cetyl alcohol

Olive Oil

Apricot kernel oil
Corn oil with homogenized crushed
   green olives
Diglycol laurate
Grapeseed oil
Lard oil with mineral oil
Mineral oil with coconut oil

Product	*Substitute or Alternative*
Olive Oil *(cont'd.)*	Peach kernel oil
	Peanut oil, destearinated
	Rice oil
	See Vegetable oils
Onyx	Petrified wood
	See Plastics
	Stone, artificial
Opacifiers	"Avicel"
	Calcium stearate
	"Polectron" 400
	Polystyrene latex
	"Rhoplex"
	Stearic acid
	Wax emulsion
Orchil Extract	*See* Dyes
Orris Root	Methyl ionone
Osage Orange Extract	*See* Dyes
Ouricuri Wax	*See* Wax
Oxalic Acid	*See* Acids
Ox-Gall	*See* Emulsifiers
	Soaps
	See Wetting agents
Oxidizers	Barium chlorate
	Barium nitrate
	Chlorine
	Chlorine dioxide
	"Dantoin"
	Potassium bromate

Product	*Substitute or Alternative*
Oxidizers *(cont'd.)*	Potassium chlorate
	Potassium nitrate
	Potassium perchlorate
	Sodium chlorcyanurate
	Sodium hypochlorite
	Sodium nitrate
	Sodium percarbonate
	Sodium peroxide
	Strontium nitrate
	Urea peroxide
Oxyquinoline Sulfate	*See* Preservatives
Ozokerite	Carnauba with amorphous paraffin wax
	See Wax
Ozone .	Chlorine
	Chlorine dioxide
	Hydrogen peroxide
Palm Oil	Glyceryl myristate
	Glyceryl oleo-stearate
	Hydrogenated vegetable oils, partially
	Mineral oil, fatty acids
	Tall oil
	Tallow
	See Vegetable oils
Palmitic Acid	*See* Fatty acids
	Stearic acid with oleic acid
Papain	Papaya juice
	Pineapple juice
Papaverine	Beta-phenylethyl-beta-methoxy-beta-phenylethyl methylamine

Product	Substitute or Alternative
Papaya Juice	Papain
Parachlormetacresol	*See* Preservatives
Parachlormetaxylenol	*See* Preservatives
Paradichlorobenzene	*See* Insecticides Naphthalene
Paraffin Oil	*See* Lubricants *See* Mineral oil
Paraffin Wax	Aluminum stearate Glyceryl tristearate Graphite Stearic acid Stearin *See* Wax
Paratoluenesulfonic Acid	Cymene sulfonic acid Phenolsulfonic acid
Paris Green	*See* Insecticides *See* Pigments
Paris White	*See* Whiting
Peach Kernel Oil	Almond oil *See* Vegetable oils
Pearl Essence	Bismuth oxychloride on mica Titanium dioxide on mica
Pectin	Agar Apple pomace Cranberries Egg albumen

Product	Substitute or Alternative
Pectin *(cont'd.)*	*See* Emulsifiers Gelatin Protein *See* Thickeners
Pentachlorethane	*See* Solvents
Pentachlorphenol	*See* Preservatives
Pentaerythritol	Mannitol Sorbitol
Pentane	Butane Ethyl alcohol
Peppermint Oil	Menthol Spearmint oil
Pepsin	Keralin
Perilla Oil	Linseed oil, boiled *See* Vegetable oil
Peroxide of Hydrogen	*See* Hydrogen peroxide
Persic Oil	Almond oil
Petrol	*See* Gasoline
Petrolatum	Microcrystalline paraffin wax with light refined mineral oil *See* Softeners
Petroleum	Shale oil
Petroleum Sulfonates	*See* Emulsifiers *See* Soaps

Product	Substitute or Alternative
Phenol	*See* Bactericides
	Cresol
	Furfural
	Oxyquinoline derivatives
	See Preservatives
	Resorcinol
Phenol Formaldehyde Resins	*See* Adhesives
	Ester gum
	See Plastics
	See Resins, natural
	See Resins, synthetic
Phenolsulfonic Acid	Benzenesulfonic acid
	Paratoluene sulfonic acid
	Xylene sulfonic acid
Phenothiazine	*See* Insecticides
Phenyl Mercuric Nitrate	*See* Preservatives
Phosphoric Acid	*See* Acids
Phthalic Anhydride	Maleic anhydride
	Phthalic acid
	Succinic acid
Phthalonitrile	*See* Insecticides
Phytosterols	Cholesterol
	Lanolin alcohols
Pigments, Inorganic	Antimony orange
	Antimony oxide
	Antimony red
	Cadmium red
	Cadmium yellow

Product	*Substitute or Alternative*
Pigments, Inorganic *(cont'd.)*	Calcium sulfate
	Carbon black
	Chrome green
	Chrome orange
	Coal, powdered
	Coal tar
	Cobalt blue
	English vermillion
	Iron blue
	Iron oxide, red
	Iron oxide, yellow
	Lampblack
	Lead chromate
	Litharge
	Lithopone
	Mercuric sulfide
	Molybdate orange
	Ocher
	Orange mineral
	Orpiment
	Red lead
	Sienna
	Tar
	Titanium oxide
	Ultramarine blue
	Umber
	Vermillion
	White lead
	Zinc oxide
	Zinc yellow
Pigments, "Organic"	Annatto
	Asphalt
	Caramel
	Carotene
	Chlorinated para red
	Cocoa

Product	Substitute or Alternative
Pigments, "Organic" *(cont'd.)*	Flushed colors
	Hansa colors
	Lignin sulfonate
	Lithol red
	Madder lake
	Molasses
	Molybdic lake
	Orange lake
	Orthonitraniline orange
	Para red
	Phospho-tungstic lake
	Phthalocyanine blue
	Phthalocyanine green
	Pitch
	Toluidine red
	Tumeric
	Vegetable colors
	Yellow lake
Pilchard Oil	Cod oil
	Manhaden oil
Pinene	Dipentene
	Pine oil
Pine·Needle Oil	Canadian fir oil
Pine Oil	*See* Insecticides
	Pinene
	See Solvents
	Sulfated tall oil
	See Wetting agents
Pine Tar	Coal tar
	Manjak
	Pitch, hardwood
	Pitch, mineral

Product	*Substitute or Alternative*
Pine Tar *(cont'd.)*	Pitch, petroleum Pitch, pine Pitch, stearin
Pipe Clay	*See* China clay Magnesium carbonate
Pitch	*See* Adhesives Asphalt Coal-tar Mineral rubber Pine tar *See* Resins Rosin
Plaster of Paris	*See* Calcium sulfate
Plastics	ABS Acrylic resins Alkyd resins Aminoplasts Aniline formaldehyde resins Asphalt with soft cumarone resins and wood flour Casein "Celluloid" Cellulose acetate Cellulose acetobutyrate Ceramics, pressed Cumarone-indene resins Ethyl cellulose Felt with asphalt or synthetic resin binder Glass, tempered Glass wool Lignin resins Melamine resins

Product	*Substitute or Alternative*
Plastics *(cont'd.)*	Methyl methacrylate
	"Neoprene"
	Phenol formaldehyde resins
	Pitch with fillers
	Polyacetal
	Polyamides
	Polycarbonate
	Polyester
	Polyethylene
	Polymethacrylate
	Polypropylene
	Polystyrene
	Polyurethane
	Polyvinyl chloride
	See Resins
	See Rubber
	SBR
	Shellac
	Sulfur with soft cumarone resin
	Tar with fillers
	"Teflon"
	Urea formaldehyde resins
	Vinylacetal resins
	Vinylbutyral resins
	Vinylidene chloride resins
	Vulcanized fiber
Plasticizers	Benzyl benzoate
	Butyl lactate
	Butyl oleate
	Camphor
	Castor oil
	Castor oil, blown
	"Clorafin"
	Diamyl phthalate
	Dibutyl phthalate
	Diglycol phthalate

Product	*Substitute or Alternative*
Plasticizers *(cont'd.)*	Dimethyl phthalate
	Diphenyl
	Diphenyl oxide
	Fatty acids
	"Flexol" plasticizer
	Glycol fatty acid esters
	Glycols
	Glycerin
	Lanolin
	Lecithin
	Mineral oil
	Oleic acid
	"Paraplex"
	Petrolatum
	Pitch
	Sorbitol
	Sulfated oils
	Tar
	Triacetin
	Tricresyl phosphate
	Triphenyl phosphate
	Vegetable oils
Platinum	Iridium
	Iron containing 42–50% nickel
	Palladium
	Rhodium
Plumbago	*See* Graphite
Polystyrene	*See* Plastics
Polyvinyl Alcohol	Gums, water dispersible
	Methyl cellulose
	Polyethylene oxide
	Polyglycols
	Synthetic resin emulsions

Product	Substitute or Alternative
Potash	See Alkalies Coconut husk ashes Fermentation residue Molasses Wood ashes
Potassium	Calcium Lithium Sodium
Potassium Bitartrate	See Cream of tartar
Potassium Bromide	Calcium bromide
Potassium Carbonate	See Alkalies
Potassium Chlorate	See Oxidizers
Potassium Cyanide	Sodium cyanide
Potassium Ethylate	Aluminum ethylate
Potassium Hydroxide	See Alkalies
Potassium Metabisulfite	Sodium bisulfite
Potassium Perchlorate	See Oxidizers
Potassium Permanganate	See Oxidizers
Potassium Silicate	See Adhesives See Alkalies
Preservatives	Acriflavin Alcohol See Bactericides Benzoic acid

Product	Substitute or Alternative
Preservatives *(cont'd.)*	BHA
	BHT
	Borax
	Boric acid
	"BTC" 2125
	t-Butyl hydroxyquinone
	Calcium propionate
	Calcium sorbate
	"Captan"
	"Chloramine" T
	Chlorothymol
	Chloroxylenol
	Copper naphthenate
	Copper sulfate
	Creosote
	Cresol
	"Dantoin"
	Diethylpyrocarbonate
	Dimethylhydantoin
	"Dowicides"
	Erythroboric acid
	Essential oils
	Ethoxyquin
	Formaldehyde
	Formic acid
	Gentian violet
	Guiac resin
	Hexachlorophene
	"Hyamine"
	Hydrogen peroxide
	Iodine
	"Kathon"
	Methyl paraben
	"Omadine"
	Organo tin oxides
	"Ottasept"
	Oxyquinoline sulfate
	Para-aminbenzoates

Product	Substitute or Alternative
Preservatives *(cont'd.)*	Paraben
	Parachlorometacresol
	Paraformaldehyde
	Phenol
	Phenylmercuric compounds
	Pine oil
	Potassium metabisulfite
	Potassium sorbate
	Propyl gallate
	Propyl paraben
	Salicylic acid
	"Santophen"
	Sodium ascorbate
	Sodium benzoate
	Sodium bisulfite
	Sodium fluoride
	Sodium metabisulfite
	Sodium nitrite
	Sodium propionate
	Sodium sulfite
	Sodium tribenzoate
	Sorbic acid
	Spices
	Stannous chloride
	Sulfur dioxide
	Thymol
	Tocopherols
	Trihydroxybutyrophenone
	Vinegar
	Wetting agents
	"Zephiran"
Propane	Acetylene
	Butane
	Natural gas
Propellents	Carbon dioxide

Product	Substitute or Alternative
Propellents *(cont'd.)*	"Freon" Hydrocarbons, chlorinated Hydrocarbons, volatile Methylene chloride
Propionic Acid	*See* Acids
Propylene Dichloride	Ethylene dichloride *See* Solvents
Propylene Glycol	*See* Solvents
Protein	Albumen Casein Corn, wheat protein Fishmeal Gelatin Gluten Soybean protein Whey proteins
Prussian Blue	*See* Iron blue
Psyllium Seed	*See* Gums, water dispersible *See* Thickeners
Pumice	*See* Abrasives
Pyrethrum	Acetylated pine oil Aliphatic thiocyanates Cucaracha *See* Insecticides Pyrethrins Rotenone
Pyridin	Aniline Bone oil

Product	Substitute or Alternative
Pyridin *(cont'd.)*	*See* Solvents
Pyrogallic Acid	Hydroquinone
Pyrogallol	Hydroquinone Naphthol, beta Pyrogallic acid Resorcinol
Pyroligneous Acid	Acetic acid
Pyrophyllite	*See* Fillers Stearite
Pyroxylin	*See* Nitrocellulose
Quartz	*See* Abrasives Garnet Silica, fused
Quercitron Bark	*See* Dyes, aniline
Quince Seed	*See* Gums, water dispersible Psyllium seed
Quinine	Pamaquine naphthoate Quinarine hydrochloride Salicin Sulfadiazin
Quinine Hydrochloride	Hydroxyethyl apocupriene Sinene
Rapeseed Oil	*See* Vegetable oils
Rayon	*See* Fibers
Rayon Flock	*See* Fillers

Product	*Substitute or Alternative*
Red Gum	*See* Yacca gum
Red Lead	Graphite Litharge *See* Pigments Zinc tetroxy chromate
Red Oil	*See* Fatty acids Fish oils Rosin oil Tall oil *See* Vegetable oils
Reducing Agents	Borane Phosphorous acid Sodium aluminum hydride Sodium borohydride Sodium hydrosulfite Sulfur dioxide Thiourea dioxide
Release Agents (Mold Releases, etc.)	"Acrawax" C "Alox" 111 "Armeen" "Armoslip" Calcium stearate Fatty acid amides Fatty acid esters Fatty acids Glycerin Glycols "Lipowax" C "Markamide" Mineral oil Silicones Soap *See* Softeners

Product	*Substitute or Alternative*
Release Agents *(cont'd.)*	Starch
	Stearic acid
	Talc
	Zinc stearate
Reodorants	Benzaldehyde
	Dichlorobenzene, para
	Nitrobenzol
Resins, Natural	*See* Adhesives
	Cellulose esters
	Ester gum
	See Resins, synthetic
Resins, Synthetic	*See* Adhesives
	Cellulose esters
	Epoxy resins
	Furfuryl lignin resin
	Hydrocarbon resins
	Melamine aldehyde resins
	Nylon
	Phenol-formaldehyde
	See Plastics
	Polyester
	Resins, natural
	See Rubber
	Silicones
	Urea-formaldehyde
Rose Oil	Butyl phenyl acetate
	Rhodinol
Rosemary Oil	Isobornyl acetate with terpinyl propionate and isoborneol
Rosin	Abietic acid
	See Adhesives

Product	Substitute or Alternative
Rosin *(cont'd.)*	Naphthenic acid
	Pitch
	See Plastics
	See Resins, natural
	See Resins, synthetic
	Tallow
Rosin Oil	Lard oil
	Mineral oil with rosin
	Tall oil
	See Vegetable oils
Rotenone	Anabasin
	See Insecticides
	Xanthone
Rottenstone	*See* Abrasives
	Tripoli
Rouge, Jeweler's	*See* Abrasives
Rubber	Acrylonitriles
	See Adhesives
	Asphalt with graphite
	EPDM
	Ethyl cellulose with castor oil
	Factice with resin and filler
	Felt with synthetic resin
	Lead oleate with carbon black
	"Neoprene"
	See Plastics
	Polyisoprene
	Polyurethane
	Polyvinyl butyral resin with 1% "Acrawax" C
	See Resins, synthetic
	Rubber dust

Product	Substitute or Alternative
Rubber *(cont'd.)*	Rubber, scrap Shellac Vinyl acetate and chloride copolymers
Rubber Cement	*See* Adhesives Blood albumen Polyvinyl acetate and copolymer emulsions *See* Resins, synthetic, solutions of *See* Rubber, synthetic solutions
Rubber, Chlorinated	Cumarone resin *See* Plastics *See* Rubber, synthetic
Rubber, Hard	Ceramics, pressed Glass, tempered *See* Plastics *See* Resins, synthetic Rubber, synthetic vulcanized Vinyl acrylic polymers Vulcanized fiber
Rubber, Sponge	Balsa wood Cellulose sponge Glass foam Mineral rubber Plastic foam *See* Plastics Sponge, natural Vulcanized vegetable oils
Rubber, Synthetic	Butyl rubber "Neoprene" "Perbunan" Polyurethane

Product	*Substitute or Alternative*
Rubber, Synthetic *(cont'd.)*	SBR "Vistanex"
Saccharic Acid	*See* Acids
Saccharin	Aspartame Glysin *See* Sugar
Saffron	Dyes Vegetable colors
Sage	Mexican oregano
Sago	*See* Adhesives *See* Gums, water dispersible Starch, sweet potato Starch waxy-sorghum *See* Thickeners
Salicylic Acid	*See* Preservatives Trichloracetic acid
Salt	Potassium chloride
Santonin	Phenothiazin
Saponin	*See* Emulsifying agents Soap bark *See* Wetting agents
Sapphire	*See* Abrasives Glass, fused hard Steel, hardened
Sardine Oil	Cod oil Rice bran oil *See* Vegetable oils

Product	Substitute or Alternative
Sebacic Acid	*See* Acids
Seed-Lac	Accroides, gum Ester gum, alcohol soluble *See* Resins, synthetic
Sequestering Agents	"Cheelox" "Chel" "Chelon" Citric acid Cream of tartar "Dequest" Diethane tritriamino *p*-acetic acid EDTA "Fostex" Gluconic acid "Hamp-ene" "Hamp-ex" "Hamp-ol" Hydroxy EDTA "Intraquest" Nitrilo acetic acid "Perma Kleer" SEQ "Seqlane" "Sequestrene" Sodium ethylene diamine acetate Sodium hexametaphosphate Tartaric acid "Versane"
Sesame Oil	Fatty glycerides *See* Vegetable oils
Shellac	Alkyd resins Batavia gum Casein Copal, alcohol soluble

Product	Substitute or Alternative

Shellac *(cont'd.)*

Ester gum with plasticizer
Gelatin
Glass with "Vinylite" coating
Glyceryl phthalate
Gum accroides
Gum kauri
Polyvinyl chloride
See Resins, synthetic
Zein

Silica

See Abrasives
See Fillers

Silica Gel

Agar
Calcium chloride
Carbon, activated
See Dessicants
Sodium alumino-silicate

Silicate of Soda

See Sodium silicate

Silicon Carbide

See Abrasives

Silicon Tetrachloride

Titanium tetrachloride

Silver Proteinate

See Preservatives

Sisal

See Fibers

Soaps

See Alkalies
Amine alcohol soaps, e.g., triethanol-
    amine oleate
Amine soaps, e.g., butylamine stearate
See Emulsifiers
Fatty acids with alkali
Lignin sulfonates
Muscovite, white

Product	Substitute or Alternative
Soaps *(cont'd.)*	Polyhydric alcohol fatty acid esters, e.g., diglycol stearate
	Saponin
	Soap bark
	Vegetable oil with alkali
	See Wetting agents
	Yucca sap
Soapstone	*See* Talc
Soda Ash	*See* Alkalies
	Sodium metasilicate
Sodium	Calcium
	Lithium
	Potassium
Sodium Abietate	*See* Soaps
Sodium Acetate	Sodium formate
Sodium Acid Sulfate	*See* Acids
Sodium Acid Sulfite	*See* Sodium bisulfite
Sodium Alginate	Agar
	Carragheen
	See Emulsifiers
	Gelatin
	See Gums, water dispersible
	See Thickeners
Sodium Alkyl Sulfate	*See* Wetting agents
Sodium Aluminate	Alum, potash
	Aluminum hydrate
	Copperas with slaked lime

Product	*Substitute or Alternative*
Sodium Arsenite	Sodium fluosilicate
Sodium Benzoate	*See* Preservatives
Sodium Bichromate	Hydrogen peroxide *See* Oxidizers
Sodium Bisulfite	Potassium metabisulfite *See* Preservatives Sodium hyposulfite
Sodium Borate	*See* Borax
Sodium Carbonate	*See* Alkalies *See* Soda ash
Sodium Caseinate	*See* Emulsifiers
Sodium Cellulose Glycolate	Methyl cellulose *See* Thickeners
Sodium Chlorate	*See* Oxidizers
Sodium Chlorite	Hydrogen peroxide *See* Oxidizers
Sodium Diacetate	Acetic acid Citric acid
Sodium Ethylate	Aluminum ethylate
Sodium Fluoride	Borax *See* Insecticides *See* Preservatives Sodium silicofluoride
Sodium Formate	Sodium acetate Sodium glycollate

Product	*Substitute or Alternative*
Sodium Glycollate	Sodium formate
Sodium Hexametaphosphate	Sodium tetraphosphate
Sodium Hydrosulfite	*See* Oxidizers
Sodium Hydroxide	*See* Alkalies
Sodium Hypochlorite	*See* Oxidizers
Sodium Hyposulfite	Sodium bisulfite Sodium chlorite
Sodium Lactate	*See* Glycerin Glycols Mineral oil
Sodium Lauryl Sulfate	*See* Wetting agents
Sodium Lignosulfonate	*See* Lignin sulfonates
Sodium Metasilicate	*See* Alkalies
Sodium Orthosilicate	*See* Alkalies
Sodium Perborate	*See* Oxidizers
Sodium Perchlorate	*See* Oxidizers
Sodium Peroxide	*See* Oxidizers
Sodium Propionate	*See* Preservatives
Sodium Pyrophosphate	Alkalies Trisodium phosphate
Sodium Resinate	Soap Sodium abietate

Product	*Substitute or Alternative*
Sodium Silicate	*See* Adhesives
	See Alkalies
	See Gums, water dispersible
	Sodium borophosphate
	Titanium sulfate
Sodium Silicofluoride	Fluorspar
	See Insecticides
	Kryolite
	Sodium fluoride
Sodium Stearate	Ammonium stearate
	Potassium stearate
	See Soaps
	See Thickeners
Sodium Sulfate	Salt
	Salt cake
Sodium Tetraborate	*See* Borax
Sodium Uranate	Cobalt oxide
Sodium Vanadate	Cobalt oxide
Softeners	"Acetulan"
	"Amerchol" 400
	"Amerlate"
	Butyl stearate
	Castor oil
	Corn syrup
	"Crodachol"
	"Crodulan"
	Degras
	"Emerest"
	Fatty acids
	Fatty alcohols

Product	*Substitute or Alternative*

Softeners *(cont'd.)*

Fatty diethanolamides
"Glucam"
Glycerin
Glyceryl fatty acid esters
Glycol fatty acid esters
Glycols
"Isopropylan"
Isopropyl myristate
Isostearic acid
Lanolin
Lecithin
"Lexate"
"Lipocal"
Mineral oil
"Minerol"
Petrolatum
Polyethylene dispersion
Quaternary ammonium compounds
"Schercomol"
Silicones
Sorbitan fatty acid esters
Sorbitol
Sulfated vegetable oils
Tall oil
Vegetable oils

Solder (Lead–Tin)

Lead, antimony, silver alloy
Lead silver alloy

Solvents

Acetone
Amyl acetate
Alcohol
Benzol
Butyl acetate
Butyl "Carbitol"
Butyl "Cellosolve"
Butyl glycollate

Product	Substitute or Alternative
Solvents *(cont'd.)*	Butyl lactate
	Butyl propionate
	"Carbitol"
	Carbon disulfide
	Carbon tetrachloride
	"Cellosolve"
	"Cellosolve" acetate
	Chlorobenzene
	Chloroform
	Cyclohexanol
	Cyclyhexanone
	p-Cymene
	"Decalin"
	Diacetone alcohol
	Dichlorodiethyl ether
	Diethyl "Carbitol"
	Diethyl carbonate
	Diethylene glycol
	Dimethyl formamide
	Dimethyl sulfoxide
	Dioxan
	Emulsions, aqueous, e.g., gum traga-canth emulsion of lemon oil and water
	Ethyl acetate
	Ethyl butyrate
	Ethyl ether
	Ethyl lactate
	Ethyl propionate
	Ethylene dichloride
	Ethylene glycol
	Ethylene glycol ethane
	"Freon"
	Furfural
	Furfurylalcohol
	Glycol diacetate
	Glycollate, diglycol
	Hexyl acetate

Product	Substitute or Alternative
Solvents *(cont'd.)*	Hydrotropic solutions, e.g., saturated water solution of sodium *p*-cymene sulfonate
	Isophorone
	Isopropyl acetate
	Isopropyl alcohol
	Isopropyl ether
	Mesityl oxide
	Methyl acetate
	Methyl acetone
	Methyl alcohol
	Methyl "Cellosolve"
	Methyl "Cellosolve" acetate
	Methyl cyclohexanol
	Methylene chloride
	Methyl ethyl ketone
	Mineral oil
	Mineral spirits (different boiling ranges)
	Nitromethane
	Octyl alcohol
	Pentachlorethane
	Pine oil
	Propylene dichloride
	Propylene glycol
	Tetrachlorethane
	Tetrahydrofurfuryl alcohol
	"Tetralin"
	Toluol
	Trichlorethylene
	Turpentine
	Water
	Xylol
Sorbitol	Corn syrup
	Dulcitol
	Glycerin

Product	*Substitute or Alternative*
Sorbitol *(cont'd.)*	Glycols
	Mannitol
	Pentaerythritol
	See Plasticizers
Spanish Oxide	*See* Pigments
Spermaceti	Cetyl palmitate
	"Spermwax"
	"Starfol" wax
	"Synaceti" 116
	See Wax, synthetic
Sperm Oil	Glyceryl monoricinoleate with thin refined mineral oil
	Lard oil, highly refined
	Neatsfoot oil
	Peanut oil, blown
	Tricresyl phosphate
	See Vegetable oils
Sponges	Cellulose (viscose) sponge
	Elastomers, foamed
	Rubbers, sponge
Squalene	Polysynlane
Starch	*See* Adhesives
	See Emulsifiers
	See Gums, water dispersible
	See Soaps
	See Thickeners
Starch Acetate	Cellulose esters
	Resins, synthetic
Stearic Acid	Abietic acid

Product	Substitute or Alternative
Stearic Acid *(cont'd.)*	Fatty acids
	Lauric acid
	Myristic acid
	Naphthenic acid
	Oxidized paraffin wax
	Palmitic acid
	Paraffin wax
	Rosin
	Tallow
Stearin	Glyceryl tristearate
	Glycol fatty acid esters
	Paraffin wax
	Zinc stearate, mineral oil
Stearyl Alcohol	Cetyl alcohol
Storax	*See* Balsam
Styrene	Acrylonitrile
Succinic Acid	*See* Acids
Sucrose	*See* Sugar
Sucrose Acetate	Cellulose esters
	Resins, synthetic
Suet	Tallow, edible
Sugar	Ammonium glycirrizinate
	Apple juice, concentrated
	Aspartamine
	Aspartane
	Calcium chloride
	Cyclamates
	Dextrin

Product	Substitute or Alternative
Sugar *(cont'd.)*	Diglycol stearate with water and saccharin
	Fructose
	Fruits, fresh, dehydrated, or juice
	Glucose
	Glycerin
	Glycols
	Gums, water dispersible
	Honey
	Invert sugar
	Lactose
	Malted barley
	Malt syrup
	Molasses
	"Nulomoline"
	Preservatives
	Saccharin
	Sorbitol
	Sorghum
	Urea
	Xylitol
Sugar Cane Wax	*See* Wax
Sugar Coloring, Burnt	*See* Caramel coloring
Sulfated Castor Oil	*See* Emulsifiers
	Naphthenic soaps
	Polyglycol fatty acid esters with or without wetting agents, e.g., nonaethylene glycol oleate
	Sulfated olive oil
	Sulfated tall oil
	Sulfated vegetable oil
	Sulfonaphthenic soaps
Sulfated Coconut Oil	Sulfated castor oil

Product	*Substitute or Alternative*
Sulfated Fatty Alcohol	*See* Emulsifiers
	See Wetting agents
Sulfated Olive Oil	Diglycol monoricinoleate
	Diglycol oleate
	See Emulsifiers
	Glyceryl monooleate
	Sulfated castor oil
Sulfated Pine Oil	Sulfated tall oil
Sulfated Red Oil	Sulfated castor oil
Sulfated Tallow	Sulfated castor oil
Sulfur	Calcium sulfide
	See Fillers
	See Insecticides
	Selenium
	Sulfur chloride
Sulfur Acid	*See* Acids
	Nitre cake
	Sodium alumino silicate
	Sodium bisulfate
	Sodium sulfate, anhydrous
Sulfur Dioxide	Chlorine
	Hydrogen peroxide
	Methyl chloride
	See Preservatives
Sumac, Silician	Acacia, shrub, domestic
	Sumac, domestic dwarf
	Tara

Product	Substitute or Alternative
Sunflower Oil	Fish oil
	Hempseed oil
	See Vegetable oils
Sunscreens	Homo methyl salicylate
	iso-Amyldimethylaminobenzoate
	p-Aminobenzoic acid and esters
Surfactants	*See* Emulsifiers
	See Wetting agents
Suspending Agents	"Avicol"
	"Bentone" Gel
	Bentonite
	"Blancol" N
	"Cab-O-Sil"
	"Carbopol"
	"Clindrol" 200-S
	CMC
	"Cosmo Wax"
	"Crodafos"
	"Cyaname" P-35
	"Dariloid"
	"Deriphat" 160C
	"Emcol" CC-9
	See Gums
	Hydroxyethyl cellulose
	"Kelco-Gel"
	"Kelcoloid"
	"Kelzan"
	"Klucel"
	Lignin sulfonates
	"Lomar" D
	"Macaloid"
	"Natrosol"
	"Polyan"
	"Reax"

Product	Substitute or Alternative
Suspending Agents *(cont'd.)*	"Syncrowax"
	"Thixcin"
	"Veegum"
	"Witcamide"
	Xanthan gum
Syrup	*See* Sugar
Tackifiers	*See* Adhesives
	Corn syrup
	See Elastomers
	Polybutenes
	Rosin esters
	Rosin oil
	See Solvents
Talc	*See* Abrasives
	Alum, potash
	Aluminum stearate
	See Fillers
	Flour, wheat
	Fuller's earth
	Graphite
	Magnesium carbonate
	Serpentine
	Soapstone
	See Wax
	Zinc stearate
Talc, Italian	Pyrophyllite
	See Talc
	Trinity (California) talc
Tallow	*See* Fatty acids
	Garbage grease
	Glyceryl oleostearate
	Hydrogenated vegetable oils

Product	Substitute or Alternative
Tallow *(cont'd.)*	Lard
	Lubricating grease
	Petrolatum
	See Soaps
	Stearin
	See Vegetable oils
	Whale oil with stearin
Tallow Oil	Menhaden oil
Tannic Acid	Alum
	Ammonium bichromate
	Formaldehyde
	Potassium bichromate
	Sodium bichromate
	See Tannin
Tannin	Dye
	Formaldehyde
	Lignin sulfonates
	Nutgalls
	Tannic acid
	Tara pods
Tapioca	*See* Adhesives
	See Gums, water dispersible
	Starch, waxy corn
	Starch, waxy sorghum
	See Thickeners
Tartar Emetic	Antimony lactate
	Antimony lactophenolate
	Sodium antimony trifluoride
	Tin ammonium chloride
Tartaric Acid	*See* Acids
Tea	Yerba mate

Product	*Substitute or Alternative*
Tea Seed Oil	Orange seed oil
Tetrachlorethane	*See* Solvents
Tetrachlorethylene	Trichlorethylene
Tetrahydrofurfuryl Alcohol	Ethyl alcohol
	See Solvents
	Trichlorethylene
Thickeners	Acrylates
	"Aerosil"
	Agar
	Albumen
	Alginates
	Aloe
	Ammonium caseinate
	Ammonium stearate
	Arrowroot
	"Avicel"
	"Barlox"
	"Bentone"
	Bentonite
	Blood, dried
	Carragheen
	Casein
	Chrondrus
	Clay
	CMC
	Collagen
	Dextrin
	Egg white
	Emulsifiers
	Fatty alkanolamides
	Flaxseed, crushed
	Gelatin
	Glucose

Product	*Substitute or Alternative*

Thickeners *(cont'd.)*

Glue
Gluten
Guar
See Gums, water dispersible
Isinglass
"Klucel"
Lanolin
Lecithin
Magnesium hydroxide (colloidal)
Magnesium trisilicate
"Methocel"
"Natrocel"
Nitrocellulose
Pectin
Polyisobutylene
Polyvinyl alcohol
Protein, fish
Protein, vegetable
Resins, synthetic
Rubbers, synthetic
See Soaps
Sodium alginate
Sodium CMC
Sodium polyacrylate
Sodium silicate
Starch
Sugar
See Suspending agents
Zein

Thinners

See Solvents

Thyme Oil

Rosemary oil

Thymol

Diisopropylmetacresol
Isopropylorthocresol
See Preservatives

Product	Substitute or Alternative
Tin Ammonium Chloride	Tartar emetic
Tin Oleate	*See* Driers Lead oleate
Tin Oxide	Antimony oxide Sodium antimonate Titanium oxide White lead Zinc oxide Zirconium oxide Zirconium silicate
Titanium Dioxide	Antimony oxide Tin oxide White lead Zinc oxide
Titanium Tetrachloride	Silicon tetrachloride
Toluol	*p*-Cymene Hydrocarbons, petroleum Hydrogenated petroleum fractions *See* Solvents
Tonka Beans	Coumarin
Triacetin	Butyl "Carbitol" "Carbitol" *See* Plasticizers Triglycollin
Tributyl Phosphate	Octyl alcohol *See* Plasticizers
Trichloracetic Acid	Salicylic acid

Product	Substitute or Alternative

Product   *Substitute or Alternative*

Trichlorethylene
    Mineral spirits
    Naphtha, petroleum (340–410°F)
    Soap with solvent
    See Solvents
    Tetrahydrofurfuryl alcohol

Tricresyl Phosphate
    Diglycol oleate
    See Plasticizers

Triethanolamine
    See Alkalies
    Alkyl amines, e.g., amylamine
    Amino alcohols, e.g., aminomethyl propanol
    See Emulsifiers

Trioxymethylene
    See Paraformaldehyde

Tripoli
    See Abrasives
    Diatomaceous earth

Trisodium Phosphate
    See Alkalies

Tritolyl Phosphate
    See Tricresyl phosphate

Tung Oil
    Castor oil, dehydrated
    Linseed oil, activated
    Linseed oil, polymerized
    Resin solutions, synthetic
    See Vegetable oils

Turmeric
    See Dyes

Turpentine
    p-Cymene
    See Solvents
    Terpinolene

Ultraviolet Absorbers
    "Amerscreen"

Product	Substitute or Alternative
Product	*Substitute or Alternative*

Ultraviolet Absorbers *(cont'd.)*

p-Aminobenzoic acid
Benzyl salicylate
"Cyasorb"
"Dipsal"
"Eastman" Inhibitor RMP
"Escalol" —
"Ferro" UV
"Giv-Tan" F
"Irgastab"
"Parsol"
"Resyn" 28-3307
"Syntase"
"Tinuvin"
"Uvinul"

Urea

Acetamide
Ammonium carbonate
Ammonium thiocyanate
Dextrin
Dicyanamid
Formamide
Hexamethylenetetramine
Melamine
See Oxidizers
Sugar
Thiourea
Titanium
Urea peroxide
Zirconium

Vanadium Pentoxide

Platinum gauze

Vegetable Oils

Animal oils
Fish oil
Glyceryl higher fatty acid esters, e.g.,
    glyceryl oleate
Mineral oil with rosin

Product	*Substitute or Alternative*
Vegetable Oils *(cont'd.)*	Rosin oil Tall oil
Venice Turpentine	Balsams Rosin oil
Vermilion Red	Antimony red *See* Pigments
Vinegar	Acetic acid Lemon juice
Vitamin E	Wheat germ
Water Absorbents	Carbon, active Cellulose CMC Diatomaceous earth Elastomers, foamed Gelatin "Klucel" "Natrocel" Silica, fumed Silica gel Starch Starch acrylates *See* Thickeners *See* Vegetable gums Viscose sponge
Water Repellents	"Aquaguard" "Aquarol" "Drilene" "Eccopel" *See* Elastomers "Emphos" D-70 "Graphsize"

Product	Substitute or Alternative
Water Repellents *(cont'd.)*	"Hydro-Pruf"
	"Masil"
	"Quso" WR
	"Raneoff"
	"Repel-O-Tex"
	"Rychem"
	"Rycowax"
	Silicones
	See Wax
Wax, Amorphous	*See* Wax
Wax, Microcrystalline	*See* Wax
Wax, Natural	Glycerides
	Lanolin wax
	Paraffin wax
	Peat wax
	Stearic acid
	Tallow
	See Wax, synthetic
Wax, Synthetic	"Acrawax"
	"Acrawax" C
	"Alfol"
	"B-Z Wax"
	"Carbowax"
	"Castor Wax"
	"Cetina"
	Chlorinated paraffin wax
	"Cyclochem" 326A
	"Dehydag" wax
	"Emerwax"
	Glyceryl tristearate
	Glycol stearates
	Hydrogenated vegetable oils
	Lignite wax

Product	*Substitute or Alternative*
Wax, Synthetic *(cont'd.)*	Monoglycerides
	"Paracin"
	Pentaerythritol stearates
	"Petrolite" WB
	Polyethylene
	"Polymekon"
	"Polyox"
	"Polywax"
	Sorbitan stearates
	"Spermwax"
	"Syncrowax"
	"Vybar"
Wetting Agents	Cresylic acid
	p-Cymene sulfonic acid, sodium salt
	Pine oil
	Polymerized glycol esters
	Quaternary ammonium compounds
	See Soap
	Sodium alkyl naphthalene sulfonate
	Sodium dioctyl sulfosuccinate
	Sulfated fatty alcohols
	Sulfated oils
	Triethanolamine
Whale Oil	Cod oil
	Mineral oils
	See Vegetable oils
Whiting	Bentonite
	See Calcium carbonate
	See Fillers
Wool Grease	Degras
	See Lanolin
	Mineral oil with limed rosin
	Petrolatum and soft soap
	Petroleum grease

Product	*Substitute or Alternative*
Xylene	*See* Xylol
Xylol	*See* Solvents
Yacca Gum	*See* Accroides, gum Resins, synthetic Shellac
Zein	Casein Protein, vegetable
Zinc Chloride	Alum Ammonium chloride Calcium chloride
Zinc Hydrosulfite	*See* Oxidizers
Zinc Oleate	*See* Driers
Zinc Oxide	Barium sulfate Titanium dioxide with talc or kaolin Whiting
Zinc Perborate	*See* Oxidizers
Zinc Peroxide	*See* Oxidizers
Zinc Phenolsulfonate	*See* Zinc sulfocarbolate
Zinc Stearate	Aluminum stearate
Zinc Sulfate	Barium sulfate, purified
Zinc Sulfocarbolate	Alum Aluminum borotartrate

Product	Substitute or Alternative
Zinc Yellow	*See* Pigments Zinc tetroxy chromate
Zirconium Oxide	Alumina Aluminum silicate Antimony oxide Lime Magnesia *See* Pigments Sodium antimonate Tin oxide Titanium dioxide Zirconium silicate
Zirconium Silicate	Chromite Dolomite, calcined Magnesite, calcined Tin oxide

Appendix

ABBREVIATIONS

amp. . . . . ampere
amp./dm^2 . . . . amperes per square decimeter
amp./sq. ft. . . . . amperes per square foot
anhydr. . . . . anhydrous
avoir. . . . . avoirdupois
Bé. . . . . Baumé
b.q. . . . . boiling point
C. . . . . Centigrade
°C. . . . . Degrees Centigrade
cc. . . . . cubic centimeter
c.d. . . . . current density
cm. . . . . centimeter
cm^3 . . . . cubic centimeter
conc. . . . . concentrated
c.p. . . . . chemically pure
cps. . . . . centipoises
cu. ft. . . . . cubic foot
cu. in. . . . . cubic inch
cwt. . . . . hundred weight
d. . . . . density
dil. . . . . dilute
dm. . . . . decimeter
dm^2 . . . . square decimeter
dr. . . . . dram
E. . . . . Engler
F. . . . . Fahrenheit
°F. . . . . Degrees Fahrenheit
f.f.c. . . . . free from chlorine
f.f.p.a. . . . . free from prussic acid
fl. dr. . . . . fluid dram
fl. oz . . . . fluid ounce
f.p. . . . . freezing point
ft. . . . . foot
ft.^{2} . . . . square foot

g. . . . . .gram
gal. . . . . .gallon
gr. . . . . .grain
hl. . . . . .Hectoliter
hr. . . . . .hour
in. . . . . .inch
kg. . . . . .kilogram
l. . . . . .liter
lb. . . . . .pound
liq. . . . . .liquid
m. . . . . .meter
min. . . . . .minim, minute
ml. . . . . .milliliter, cubic centimer
mm. . . . . .millimeter
m.p. . . . . .melting point
N. . . . . .Normal
N.F. . . . . .National Formulary
oz. . . . . .ounce
pH . . . . .hydrogen-ion concentration
p.p.m. . . . . .parts per million
pt. . . . . .pint
pwt. . . . . .pennyweight
q.s. . . . . .a quantity sufficient to make
qt. . . . . .quart
r.p.m. . . . . .revolutions per minute
S.A.E. . . . . .Society of Automotive Engineers
sec. . . . . .second
sp. . . . . .spirits
sp. gr. . . . . .specific gravity
sq. dm. . . . . .square decimeter
tech. . . . . .technical
tinc. . . . . .tincture
tr. . . . . .tincture
Tw. . . . . .Twaddell
U.S.P. . . . . .United States Pharmacopoeia
v. . . . . .volt
visc. . . . . .viscosity
vol. . . . . .volume
wt. . . . . .weight
x. . . . . .extra

COMMON AND CHEMICAL NAMES

A

Acacia gum—gum arabic
Acetate of lime—calcium acetate
Acetic ether—ethyl acetate
Acetin—glyceryl monoacetate
Acetyl salicylic acid—aspirin
Acetylene tetrachloride—tetrachlorethane
Adeps lanae—lanolin
Alcohol—ethyl alcohol
Aldehyde—acetaldehyde
Alumina—aluminum oxide
Alum—potassium aluminum sulfate
Alundum—fused aluminum oxide
Ammonia, aqua—ammonium hydroxide
Aniline oil—aniline
Animal charcoal—bone black
Aqua fortis—nitric acid
Aqua regia—mixture of nitric and hydrochloric acid
Argols—crude cream of tartar
Arsenic, red—arsenic disulfide
Asphaltum—mineral pitch

B

Baking soda—sodium bicarbonate
Banana oil—amyl acetate
Barium white—barium sulfate
Baryta—barium oxide
Barytes—barium sulfate, natural
Bauxite—aluminum oxide, hydrated
Benzene—benzol
Black boy gum—accroides gum
Black hypo—lead thiosulfate, impure
Blanc fixe—barium sulfate, precipitated
Bleaching powder—calcium hypochlorite
Blue lead, sublimed—basic lead sulfate
Blue stone
Blue vitriol }—copper sulfate
Boiled oil—linseed oil, boiled
Bone black—animal charcoal

Boracic acid—boric acid
Borax—sodium tetraborate
Brazil wax—carnauba wax
Brimstone—sulfur
British gum—dextrin
Bromo acid—tetrabromfluorescein
Burnt sugar coloring—caramel color
Butanol—butyl alcohol
Butter color—annatto
Butter of antimony—antimony chloride
Butyric ether—ethyl butyrate

C

Cadmium yellow—cadmium sulfide
Calcium phosphate—calcium phosphate, monobasic
Calomel—mercurous chloride
Cane sugar—sucrose
Carborundum—silicon carbide
Capsicum—red pepper
Carbolic acid—phenol
Carragheen—Irish moss
Catechu—cutch
Caustic potash—potassium hydroxide
Caustic soda—sodium hydroxide
Ceresin wax—ozokerite and paraffin wax mixture
Chalk—calcium carbonate
China clay—aluminum silicate
China wood oil—tung oil
Chinese wax—insect wax
Chloride of lime—calcium hypochlorite
Cholestrin—cholesterol
Chrome green—lead or zinc chromate or ferric ferrocyanide
Chrome yellow—lead chromate
Cinnabar—mercuric sulfide
Citronella oil—verbena oil
Cocoa butter—theobroma oil
Cognac oil—oenanthic ether
Colloidal clay—bentonite
Collodion—nitrocellulose solution
Cologne spirits—ethyl alcohol, pure

Colophony—rosin, pine resin
Columbian spirits—methyl alcohol
Colza oil—rapeseed oil
Copper aceto-arsenite—Paris green
Copper arsenite—Scheele's green
Corn sugar—dextrose
Corn syrup—glucose, mixture of dextrin and dextrose and maltose
Corrosive sublimate—mercuric chloride
Corundum—aluminum oxide
Cream of tartar—potassium bitartrate
Cresol—cresylic acid
Crude oil—petroleum (crude)
Cyanamid—calcium cyanamide

D

Dead oil—creosote oil
Degras—wool grease
Dekalin—decahydronaphthalene
Dextrose—corn sugar
Disodium phosphate—sodium phosphate, dibasic
Dope—pyroxylin "solution"
Dutch liquid—ethylene chloride

E

Earth, infusorial—diatomaceous earth
Earth wax—ozokerite, mineral wax
Egg oil—egg yolk
Elaterite—mineral rubber
Epsom salts—magnesium sulfate
Emery—impure aluminum oxide
Ester gum—glycerol ester of rosin
Ether—ethyl ether

F

Fir balsam—Canada balsam
Fixed white—barium sulfate
Flaxseed—linseed
Flea seed—psyllium
Fluorspar—calcium fluoride
Fool's gold—iron pyrite

Formalin—formaldehyde (40% solution)
French chalk, talc—magnesium silicate
Fuchsine—magenta
Fusel oil—amyl alcohol, crude
Fuller's earth—aluminum silicate, hydrous

G

Galena—lead sulfide
Gasoline—petroleum spirit
Glance pitch—manjak
Glass, water—sodium silicate
Glauber salt—sodium sulfate ($10H_2O$)
Glycerin—glycerol
Glucose—corn syrup
Glycol—ethylene glycol
Graphite—plumbago
Green soap—soft soap
Green vitriol—ferrous sulfate
Grain alcohol—ethyl alcohol
Ground nut oil (Arachis oil)—peanut oil
Gum lac—shellac
Guncotton—nitrocellulose
Gypsum—calcium sulfate
Gugnets' green—chromium oxide, hydrated

H

Heavy spar—barium sulfate
Hematite—iron oxide
Hexalin—cyclohexanol
Hexamine—hexamethylenetetramine
Hydrosulfite—sodium hydrosulfite
Hypo—sodium thiosulfate

I

Ichthyol—ammonium sulfo-ichthylate
Indene—*p*-cumarone
Indian gum—karaya, gum
Indian red—ferric oxide
Insect flowers (powdered)—pyrethrum
Isinglass—pure fish gelatin
Italian red—iron oxide (red)
Ivory black—bone black

K

Kauri gum—copal, gum
Kaolin—aluminum silicate
Kieselguhr—diatomaceous earth

L

Lanolin—purified wool grease
Lanum—lanolin
Lead chromate—chrome yellow
Lead sulfide, basic—white lead, sublimed
Lemon, salts of—potassium binoxalate
Lemon yellow—barium chromate
Licorice—glycyrrhiza
Ligroin, light—petroleum ether
Lime, dry—calcium oxide
Lime, slaked—calcium hydroxide
Limestone—calcium carbonate
Litharge—lead monoxide
Liver of sulfur—potassium sulfide
Lithopone—zinc sulfide and barium sulfate
Lunar caustic—silver nitrate
Lye—sodium hydroxide
Lysol—cresol soap solution

M

Magnesia—magensia oxide
Magnesite—magnesium carbonate, natural
Magnesium silicate—talc
Maize oil—corn oil
Malt sugar—maltose
Metol—methyl-*p*-aminophenol sulfate
Methanol—methyl alcohol
Microcosmic salt—sodium ammonium phosphate
Milk sugar—lactose
Mineral wax—ozokerite
Mineral pitch—asphalt
Minium—lead oxide (red)
Mirbane oil—nitrobenzol
Muriatic acid—hydrochloric acid
Myrtle wax—bayberry wax

N

Naphtha (petroleum)—petroleum distillate

Naphtha, solvent—coal tar naphtha
Naples yellow—lead antimonate
Nickel salts, double—nickel ammonium sulfate
Nickel salts, single—nickel sulfate
Niter—potassium nitrate
Nitrous ether—ethyl nitrate

O

Oil of bitter almond—benzaldehyde (from bitter almond nut)
Oil of mirbane—nitrobenzol
Oil of mustard—allyl isothiocyanate
Oil of wintergreen—methyl salicylate
Oleic acid—red oil
Olein—glyceryl trioleate (natural)
Oleum—sulfuric acid (fuming)
Orange mineral—lead oxide (orange red)
Orpiment—arsenous sulfide (yellow)

P

Paraffin oil—mineral oil, petrolatum liquid
Paris white—calcium carbonate
Paris blue—ferric ferrocyanide
Pearl ash—potassium carbonate
Petrol—gasoline
Petrolatum—petroleum jelly
Plaster of Paris—calcium sulfate
Plumbago—graphite
Prussian blue—ferric ferrocyanide
Prussiate of potash, red—potassium ferricyanide
Prussiate of potash, yellow—potassium ferrocyanide
Prussic acid—hydrocyanic acid
Pyrethrum—insect flowers (powdered)
Pyroligneous acid—wood vinegar (acetic acid)
Pyrolusite—manganese dioxide
Pyroxylin—nitrocellulose

Q

Quicklime—calcium oxide
Quicksilver—mercury
Quinol—hydroquinone

R

Red lead—lead tetroxide

Red oil—oleic acid
Red oxide—ferric oxide (red)
Rochelle salt—potassium sodium tartrate
Rosin—pine resin, colophony
Rottonstone—tripoli
Rouge—ferric oxide

S

Sal ammoniac—ammonium chloride
Satin white—reaction product of hydrated lime and alum
Salt—sodium chloride
Saltpeter—potassium nitrate
Salts of vitriol—zinc sulfide
Scale wax—paraffin wax (low melting point)
Silica—silicon dioxide
Slaked lime—calcium hydroxide
Sod oil—degras
Soda ash—sodium carbonate
Soda (washing)—sodium carbonate ($10H_2O$), sal soda
Sodium bisulfite—sodium acid sulfite
Sodium phosphate, dibasic—disodium phosphate
Soft soap—potash soap
Stearin—stearic acid
Sublimed lead—lead sulfate, basic
Sucrose—cane sugar, beet sugar
Sugar of lead—lead acetate
Sulfuric ether—ethyl ether

T

Talc—magnesium silicate
Tartar emetic—antimony potassium tartrate
Tetralin—tetrahydronaphthalene
Theobroma oil—cocoa butter
Titanium dioxide—titanium oxide
Toluol—toluene
Triacetin—glycerol triacetate
Tripoli—natural amporphous silica, not diatomaceous
Tripolite—diatomaceous earth
Train oil—whale oil
Trinitrophenol—picric acid
Turkey red oil—castor oil, sulfonated

V

Venetian red—ferric oxide
Vermilion—red mercuric sulfide
Verdigris—copper acetate, basic
Vitriol—sulfuric acid

W

Water glass—sodium silicate
White arsenic—arsenic trioxide
White bole—China clay, kaolin
White lead—lead carbonate, basic
White lead, sublimed—lead sulfide, basic
White metal—Babbitt metal
White wax—beeswax, bleached
Whiting—calcium carbonate
Witherite—barium carbonate, natural
Wintergreen oil, synthetic—methyl salicylate
Wood alcohol—methyl alcohol

X

Xylol—xylene

Y

Yacca gum—accroides gum

Z

Zinc white—zinc oxide
Zinc yellow—potassium zinc chromate

SOME INCOMPATIBLE CHEMICALS

Some Incompatible Chemicals

The substances in the left-hand column must be stored and handled so that they cannot come into contact with the substances in the right-hand column.

Alkaline and alkaline-earth metals, such as sodium, potassium, lithium, magnesium, calcium, aluminum	Carbon dioxide, carbon tetrachloride, and other chlorinated hydrocarbons. (Also prohibit water, foam, and dry chemical on fires involving these metals.)
Acetic acid	Chromic acid, nitric acid, hydroxyl containing compounds, ethylene glycol, perchloric acid, peroxides, and permanganates.
Acetone	Concentrated nitric and sulfuric acid mixtures.
Acetylene	Chlorine, bromine, copper, silver, fluorine, and mercury.
Ammonia (anhydr)	Mercury, chlorine, calcium hypochlorite, iodine, bromine, and hydrogen fluoride.
Ammonium nitrate	Acids, metal powders, flammable liquids, chlorates, nitrites, sulfur, finely divided organics or combustibles.
Aniline	Nitric acid, hydrogen peroxide.
Bromine	Ammonia, acetylene, butadiene, butane and other petroleum gases, sodium carbide, turpentine, and finely divided metals.
Calcium carbide	Water (See also acetylene.)
Calcium oxide	Water.
Carbon, activated	Calcium hypochlorite.
Copper	Acetylene, hydrogen peroxide.
Chlorates	Ammonium salts, acids, metal powders, sulfur, finely divided organics or combustibles.

Chromic acid	Acetic acid, naphthalene, camphor, glycerol, turpentine, alcohol, and other flammable liquids.
Chlorine	Ammonia, acetylene, butadiene, butane and other petroleum gases, hydrogen, sodium carbide, turpentine, benzene, and finely divided metals.
Chlorine dioxide	Ammonia, methane, phosphine, and hydrogen sulfide.
Fluorine	Isolate from everything.
Hydrocyanic acid	Nitric acid, alkalis.
Hydrogen peroxide	Copper, chromium, iron, most metals or their salts, any flammable liquid, combustible aniline, nitromethane.
Hydrofluoric acid (anhydr.) (hydrogen fluoride)	Aqueous or anhydrous ammonia.
Hydrogen sulfide	Fuming nitric acid, oxidizing gases.
Hydrocarbons (benzene, butane, propane, gasoline, turpentine, etc.)	Fluorine, chlorine, bromine, chromic acid, sodium peroxide.
Iodine	Acetylene, anhydrous or aqueous ammonia.
Mercury	Acetylene, fluminic acid, ammonia.
Nitric acid (conc.)	Acetic acid, aniline, chromic acid, hydrocyanic acid, hydrogen sulfide, flammable liquids, flammable gases, and nitritable substances.
Nitroparaffins	Inorganic bases.
Oxygen	Oils, grease, hydrogen, flammable liquids, solids or gases.
Oxalic acid	Silver, mercury.
Perchloric acid	Acetic anhydride, bismuth and its alloys, alcohol, paper, wood, grease, oils.
Peroxides, organic	Organic or mineral acids; avoid friction.

Phosphorus (white)	Air, oxygen.
Potassium chlorate	Acids (see also chlorate.)
Potassium perchlorates	Acids (see also perchloric acid.)
Potassium permanganate	Glycerol, ethylene glycol, benzaldehyde, sulfuric acid.
Silver	Acetylene, oxalic acid, tartaric acid, fulminic acid, ammonium compounds.
Sodium	See alkaline metals.
Sodium nitrate	Ammonium nitrate and other ammonium salts.
Sodium oxide	Water.
Sodium peroxide	Any oxidizable substance, such as ethanol, methanol, glacial acetic acid, acetic anhydride, benzaldehyde, carbon disulfide, glycerol, ethylene glycol, ethyl acetate, methyl acetate, and furfural.
Sulfuric acid	Chlorates, perchlorates, permanganates.
Zirconium	Prohibit water, carbon tetrachloride, foam, and dry chemical on zirconium fires.

COMMON HAZARDOUS CHEMICALS

Name	Usual Shipping Container	Fire Hazard	Life Hazard	Storage	Fire-Fighting Phases	Remarks
Acetic acid (glacial)	Glass carboys and barrels	Dangerous in contact with chromic acid, sodium peroxide, or nitric acid; yields moderately flammable vapors above flash point 104° F.	May cause painful burns of skin	Safeguard against mechanical injury. Isolate from oxidizing materials as noted under Fire Hazard	Extinguishing agent, water	Expands on solidification and may burst container unless kept at a temperature above 16°C. (60.8°F.)
Acetone	Carboys, steel drums, tank cars	A volatile liquid. Gives off vapors which form with air flammable and explosive mixtures. Flash point -16°C. (3°F.). Explosive range 2.55% (3). The ignition to 12.8% (upward propagation) temperature is comparatively high, being within the range 538° to 566°C. 1000° to 1050° (F.), (II). The	Toxicity of a comparative order (7) low	Safeguard containers against mechanical injury. Only electrical equipment of the explosion-proof type, Group D classification, permitted in atmospheres containing acetone vapor in flammable proportions	Lighter than water (sp. gr. 0.792) but soluble in it in all proportions. Water, particularly in form of spray, is best extinguisher. Carbon dioxide may also be used. Automatic sprinkler systems or total-flooding carbon dioxide systems may be employed for protection in storage rooms	

Note: These columns are informative only; it is not considered necessary that the material be kept or stored only in the containers as listed, nor that each package be labeled. The requirements in the table on the storage of containers refer to chemicals in usual containers, and are not intended to apply to small bottles of chemicals such as are found in drug stores and chemical laboratories.

Name	Usual Shipping Container	Fire Hazard	Life Hazard	Storage	Fire-Fighting Phases	Remarks
Aluminum dust	Barrels or boxes	vapors are heavier than air (vapor density 2), Fire hazard slightly less than that of gasoline		Keep in dry place. In case of fire do not use water; it may cause an explosion	Smother with sand, ashes, or rock dust. Do not use water, which may cause explosion	
Aluminum resinate	Wooden barrels	Forms flammable and explosive mixtures with air				
		Combustible		Storage should be ventilated and safeguarded as for oil storage building		
Ammonium perchlorate	Wooden barrels or kegs and glass bottles	Oxidizing material. May explode in a fire. Hazard classes with potassium chlorate		Safeguard against mechanical injury. Isolate from mineral acids, also from combustibles		
Anhydrous ammonia	Steel cylinders or steel tank cars	Gas density 0.60 (air = 1). Not flammable in air except in comparatively high concentration, which is seldom encountered under practical conditions, the low limit of the flammable or explosive range being about 15 to 16% and the upper limit about 25 to 26% by volume (horizontal propagation, 8, 14). Presence of oil will increase the	Irritant. An outstandingly serious effect produced by ammonia in concentrations of the order of 0.5% by volume for duration of exposure of the order of 0.5 hour is blindness (12). A concentration of 0.30% of ammonia in air for duration of exposure of the order of o.5 to 1 hour. according to Lehmann, does not cause	Safeguard against mechanical injury and excessive heating of cylinders or tanks. Fire-resistive storage recommended. In combustible buildings or if near combustible storage, sprinkled storage recommended. Isolate from other chemicals, particularly chlorine, bromine, iodine, and mineral acids	Soluble in water. Hose streams comparatively effective in removing gas from atmosphere	

	Container	Hazard	Serious effects	Storage	Fire extinguishing	Remarks
Antimony pentasulfide (golden antimony sulfide) Sb_2S_5	Fiber drums or tins	hazard. Ammonia aqua does not burn. Combustible. Readily ignited by small flame. Hazardous in contact with oxidizing material. Yields flammable hydrogen sulfide on contact with mineral acid	Gaseous products of combustion contain sulfur dioxide and are irritating and corrosive	Safeguard against mechanical injury. Isolate from acids, chlorates, nitrates, and other oxidizing agents	Practically insoluble in water. Use water	Used in manufacture of matches, ammunition, fireworks, and of certain rubber compounds
Barium chlorate	Wooden boxes, barrels, or kegs	Oxidizing material. Hazard classes with potassium chlorate		Isolate	See potassium chlorate	
Barium nitarte	Wooden boxes; barrels	Oxidizing material. Hazard in class with sodium nitrate	Soluble in water. Poisonous when taken internally	Do not store with combustible materials	See sodium nitrate	
Barium peroxide	Tihgtly closed metal containers packed in wooden boxes or in barrels; or in bulk in metal barrels or drums	Oxidizing material. Hazard in class with sodium peroxide		Do not store with combustible materials	Smother with sand, ashes, or rock dust. Do not use water	
Benzoyl peroxide (dry) granular or powdered (wet)	Dry granular material in individual 1-lb. containers inside wooden boxes; finely powdered material shipped wet (30% water by weight) in glass containers tightly placed inside sealed metal containers in wooden boxes or in aluminum drums	Highly flammable in the dry state. Strong supporter of combustion. Do not subject dry material to heat of friction or grinding	Dust irritating to eyes and lungs. Use goggles and dust respirator in dusty atmospheres	Store in cool ventilated place. Powder should be stored with not less than 30% water by weight. Keep away from all sources of heat and separate from all combustible materials and acids	Water, carbon dioxide, foam, sand, soda ash, or rock dust may be used as extinguishing agents	Not miscible with water

Name	Usual Shipping Container	Fire Hazard	Life Hazard	Storage	Fire-Fighting Phases	Remarks
Bleaching powder, calcium hypochlorite, chlorinated lime, chloride of lime (incorrect name)	Air-tight tin containers, wooden barrels, and steel drums	Notcombustible but evolves chlorine and at higher temperatures oxygen. With acids or moisture evolves chlorine freely at ordinary temperatures. See chlorine	Corrosive. Irritating to skin, eyes, and lungs. See chlordine	Store in cool, dry, well-ventilated place away from combustibles. See chlorine. Rupture of drums containing bleaching powder, particularly if the chlorine content is high, may result from exposure to heat.	Fires where compound is present may be fought with water, preferably spray. Protect eyes and skin, using gas mask of type approved by Bureau of Mines	Encountered in paper, textile, disinfectant, and alkali industries; also where water-purification processes are employed
Borneol	Barrels, kegs, boxes, and tins	Combustible. Hazard similar to camphor		Store in well-ventilated compartment or building	See camphor	
Bromine	Glass bottles; earthen jugs	Causes oxidizing effect, resulting in heating and may cause fire when in contact with organic material	Corrosive; at ordinary temperatures gives off poisonous suffocating vapor	Isolate; safeguard against mechanical injury		Bottles should be surrounded by incombustible packing
Bronze dust	Barrels or boxes	When aluminum is present forms flammable and explosive mixtures with air. Composition usually free from aluminum			Smother with sand or ashes	Bronze dust free from aluminum not considered dangerous
Butane	Steel cylinders	Flammable gas under pressure. Classes with gasoline vapor in fire hazard		Safeguard against mechanical injury		Keep cool
Calcium carbide	Iron drums and tin cans	Gives off acetylene gas on contact with water or moisture	Serious under fire conditions	Store in dry, well-ventilated place in accordance with N.F.P.A. Standards	Smother with sand or ashes. Do not use water	
Calcium oxide	Wooden barrels and bags	Heats upon contact with water or moisture and		Isolate; store in dry place away from water or		

Name	Container		Properties	Toxicity	Storage	Fire extinguishing	Handling
Camphene	Tins	moisture	may cause ignition of organic material. Swells when moist and may burst container		When heated gives off flammable vapors. Classes with turpentine	Isolate; keep in unheated compartment away from fire or heat	Smother with sand or ashes. Avoid water
Camphor	Tins and wooden kegs		Flammable; gives off flammable vapors when heated which may form explosive mixture with air. Flash point 180°F.		Detach from other storage. Keep in well-ventilated room remote from fire	Smother with sand or ashes. Chemical streams. Avoid water	
Carbolic acid (see phenol)							
Carbon disulfide	Small glass, earthenware, or metal containers packed in outside barrels or boxes (See I.C.C. Regulations). Steel drums, steel tank cars		Highly volatile liquid with offensive odor, giving off even at comparatively low temperatures vapors which form with air flammable and explosive mixtures. Flash point, $-30°C$. $(-22°F.)$. (11). Flammable range 1 to 50% (7) (upward propagation). Ignition temperature is dangerously low, about 100° to 106°C. (212° to 223°F.). It is endothermic, and vapor may be ignited by heavy blow. Vapors are heavier than	Toxic. 3200 to 3850 parts of vapor per million (0.32 to 0.385%) by volume) may cause dangerous illness in 0.5 to 1 hour (5). Direct contact with skin should be avoided. Products of combustion contain sulfur dioxide, which in concentrations of 0.2% by volume in air may cause serious injury in 0.5 hour or less (12). Often poisonous carbon monoxide is present in products of combustion	Isolate and safeguard containers against mechanical injury and metallic blows, and keep in unheated compartment away from sunlight and any source of ignition, including electric lighting fixtures and other electrical equipment. Storage tanks should be constructed over concrete basins containing water, and carbon disulfide kept blanketed with water or inert gas at all times	Heavier than water (sp. gr. 1.29) and sparingly soluble in it. Use sand, carbon dioxide, or other inert gas as extinguishing agents. Cooling and blanketing action of water may be utilized in case of fires in metal containers or tanks. Total-flooding carbon dioxide systems may be employed for protection in storage rooms. Foam not effective. Do not use carbon tetrachloride. Use of gas masks or oxygen helmets of type approved	Carbon disulfide should never be transferred by means of air. Use inert gas, water, or pump. Use a wood measuring stick for measuring contents of storage tanks or tank cars. Tank cars when being loaded or unloaded should be well grounded. Do not dispose of carbon disulfide by pouring it on the ground. Provide a safe place for burning it.

Name	Usual Shipping Container	Fire Hazard	Life Hazard	Storage	Fire-Fitting Phases	Remarks
		air (vapor density 2.62) and may travel considerable distance to source of ignition and flash back. More hazardous than gasoline			for purpose by U.S. Bureau of Mines recommended	
Charcoal (wood)	Boxes, barrels, bags, or bulk	Spontaneously ignitable when freshly calcined and exposed to air, or when wet; hazardous when freshly ground and tightly packed	Danger from carbon monoxide poisoning during burning unless adequate ventilation is provided	Isolate; prevent accumulations; ventilate well; make daily inspections	Use water, completely extinguishing fire, after which storage pile should be moved	
Chinese wax	Burlap bags and wooden barrels	Combustilbe				
Chlorine	Steel cylinders and tank cars	Not combustilbe in air but reacts chemically with many common substances and may cause fire or explosion when in contact with them. See remarks under Storage	Corrosive. Irritating to eyes and mucous membrane. Toxic. 0.004 to 0.006% by volume in air causes illness in 0.5 to 1 hour (6)	Detach from other storage Isolate from turpentine, ether, ammonia gas, illuminating gas, hydrocarbons, hydrogen, and finely divided metals. Safeguard against mechanical injury of containers	Use gas masks on entering atmospheres containing chlorine gas. If, however, concentration is high, or there is doubt as to degree of concentration, use oxygen helmet of type approved for such use by U.S. Bureau of Mines	Dangerous to neutralize chlorine in a room with ammonia
Chromium trioxide or chromium anhydride, CrO_3 (ofen called "chromic acid")	Iron drums and glass bottles	Oxidizing material; will ignite on contact with acetic acid and alcohol. Chars organic material such as wood, sawdust, paper, or cotton, and may cause ig-	Irriritatng to skin. Poisonous	Isolate	Use water, completely extinguishing fire, after which storage pile should be removed	Used in chromium plating, in electric batteries, and in photography

Name	Container	Hazard	Use / Health	Storage	Remarks
Cobaltous nitrate	Wooden barrels	nition. Combustible material in presence of chromium trioxide when ignited burns with great intensity. May cause explosion in fire. Oxidizing material. Classes with sodium nitrate		See sodium nitrate	
Colophony	Barrels	Combustible; gives off flammable vapors when heated		Ventilate storage; avoid dust; keep away from fire or heat	
Copper nitrate	Wooden barrels and kegs	Oxidizing material. Hazard classes with sodium nitrate	Poisonous when taken internally. Soluble in water	Safeguard against mechanical injury; isolate. See sodium nitrate	
Cyclopropane	Steel cylinders	Highly flammable gas. Forms flammable and explosive mixtures with air or oxygen. Explosive range 2.40 to 10.4% (upward propagation) (4)	Anesthetic	Isolate from oxygen cylinders and store in cool, well ventilated storeroom	Use water to cool cylinders not on fire. If gas is burning at valves or safety releases, usually best course to follow is not to disturb or attempt to extinguish flame. To do so will cause release of unburned gas and quickly create high dangerous explosive atmospheres. If cylinder is mounted on an esthetic machine or truck, it may be possible to move it to a safe place. Carbon Only electrical equipment of the explosion-proof type, Group C classification, permitted in atmospheres containing cyclopropane in flammable proportions. Group C classification for cyclopropane tentative pending further tests

Name	Usual Shipping Container	Fire Hazard	Life Hazard	Storage	Fire-Fighting Phases	Remarks
Didymium nitrate	Wooden kegs	Oxidizing material. Classes with sodium nitrate		Isolate. See sodium nitrate	dioxide or carbon tetrachloride are best extinguishing agents	
Dioxan	Glass bottle, metal cans, and metal drums	Moderately volatile flammable liquid. Flash point 54°F. Explosive range 1.97 to 22.25% (upward propagation) (14). Vapors heavier than air (vapor density 3.03). Capable of forming peroxides under certain conditions, and may be danger of explosion if redistilled, unless certain precautions are taken	Irritant and toci in high concentrations	Isolate and safeguard against mechanical injury	Water best extinguishing agent	Slightly heavier than and completely soluble in water (sp. gr. 1.03)
Ether, ethyl	Glass bottles or tin cans in boxes, steel drums	Highly volatile liquid, giving off even at comparatively low temperatures vapors which form with air or oxygen and flammable and explosive mixtures. Explosive range 1.85 to 36.5% (upward propagation) (2). Ignition temperature	Anesthetic	Safeguard containers against mechanical injury. Isolate and keep in unheated compartment away from sunlight and any source of ignition. Only electrical equipment of explosion-proof type, Group C classification, per-	Lighter than water (sp. gr. 0.7135). Soluble in about ten times its own volume of water. Water may be utilized only to cool metal containers. Best extinguishing agents are carbon dioxide and sand, also carbon tetrachloride in	See National Board of Fire Underwriter's Recommended Safeguards for the Installation and Operation of Anaesthetical Apparatus Employing Combustible Anaesthetics

			is comparatively low, approximately 180°C. (356°F.) (11). Spontaneously explosive peroxides sometimes form on long standing or exposure in bottles to sunlight. Vapors heavier than air (vapor density 2.6) and may travel considerable distance to source of ignition and flash back. More hazardous than gasoline	mitted in atmospheres containing ether vapor in flammable proportions	case of fires involving limited amounts of ether. Total-flooding carbon dioxide systems may be employed for protection in storage rooms
Ethylene	Steel cylinders	Anesthetic	Highly flammable gas. Forms flammable and explosive mixtures with air or oxygen. Explosive range 2.75 to 28.6% (upward propagation) (4). Ignition temperature about 450°C. (842°F.) (10). Slightly lighter than air (density 0.97) Gas is spontaneously explosive in sunlight with chlorine	Isolate from oxygen cylinders and store in a cool, well-ventilated storeroom	Use water to cool cylinders not on fire. If gas is burning at valves or safety releases, usually best course to follow is not to disturb or attempt to extinguish flame. To do so will cause release of unburned gas and quickly create highly dangerous explosive atmospheres. If cylinder is mounted on anesthetic machine or truck, it may be possible to move it to safe place. In closed storeroom car-
					Only electrical equipment of explosion - proof type, Group C classification, permitted in atmospheres containing ethylene in flammable proportions

Name	Usual Shipping Container	Fire Hazard	Life Hazard	Storage	Fire-Fighting Phases	Remarks
Ferric nitrate	Wooden barrels	Oxidizing material		Isolate. Safeguard against mechanical injury	bon dioxide or cabon tetrachloride are best extinguishing agents	
Formic acid	Barrels and carboys	Flammable; gives off flammable vapors which may form explosive mixtures with air	Corrosive; has caustic effect on the skin	Safeguard against mechanical injury		
Fulminate of mercury		High explosive (primary class)		Explosives restrictions		
Fulminate of silver		High explosive (primary class)		Explosives restrictions		
Hydrichloric acid (muriatic acid)	Tank cars (rubber-lined), carboys, and bottles	Not combustible (in air) but if allowed to come in contact with common metals hydrogen is evolved, which may form explosive mixtures with air	Aqueous solution is corrosive, irritating, and poisonous. Fumes corrosive and irritating to mucous membranes	Safeguard containers against mechanical injury. Keep away from oxidizing agents, particularly nitric acid and chlorates. Avoid contact by leakage or otherwise with all common metals	Use water or chemically basic substances such as soda ash or slaked lime	
Hydrcyanic acid (prussic acid)	Cylinders, or when completely absorbed in inert material in metal cans with outside wooden boxes	Forms flammable and explosive mixtures with air. Explosive range 6 (13) to about 40% by volume (horizontal propagation)	Poisonnous. Few breaths may cause unconsciousness and death. Avoid contact with skin	Isolate. Keep away from any source of heat. Safeguard containers against mechanical injury	Gas is slightly lighter than air. Soluble in water. Water is best extinguisher. When entering premises where used or stored during fire, use oxygen helmet or gas mask equipped with canister of approved type approved by Bureau of	Concentrations of gas ordinarily employed for fumigation (1% or less) are considerably below the lower limit of flammability (6% by volume). Some methods of fumigation, however, are employed which temporarily

Substance	Containers	Properties	Hazards / Health	Precautions	Fire Fighting / Protection	Remarks
(continued from previous entry)					Mines for hydrocyanic acid	yield flammable mixtures even though final concentration is low.
Hydrofluoric acid (HF)	Aqueous solution in lead carboys and wax or gutta percha bottles	Colorless, volatile liquid. Not combustible but reacts with glass and most substances, platinum being an exception. Aqueous solution also attacks glass and several metals	Acid and its vapors highly toxic and irritating to skin, eyes, and respiratory tract. Fumes produced by contact with ammonia and many metals poisonous. May be neutralized with chalk. Bicarbonate of soda solution may be immediately applied to burns as first-aid and used as gargle	Isolate. Ventillate. Safeguard against mechanical injury	Use water in case of fires involving hydrofluoric acid. Use oxygen helmet of type approved for such use by U.S. Bureau of Mines on entering atmospheres known to contain hydrofluoric acid vapors	Encountered in glass works and chemical laboratories. Used to remove sand from castings and in manufacture of filter paper. Vapors have been known to cause serious corrosion of sprinkler piping and heads
Hydrofluosilicic acid	Lead carboys, hard rubber or paraffin bottles	None	Corrosive	Safeguard against mechanical injury	Use water	
Hydrogen peroxide (27.5% by weight)	Glass carboys, aluminum drums, aluminum tank cars (all containers must be vented)	Oxidizing liquid. May cause ignition of combustible material if left standing in contact with it. May decompose violently if contaminated with iron, copper, chromium, and most metals or their salts	Prolonged exposure to vapor irritating to eyes and lungs. Causes skin irritation	Use goggles to protect eyes from splash. Store in cool place in ventilated containers remote from combustible material and catalytic metals such as iron, copper, chromium		
Hydrogen sulfide (sulfuretted hydrogen)	Steel cylinders	Flammable gas. Forms flammable and explosive mixture with air or oxy-	Toxic; 0.05 to 0.07% by volume in air causes dangerous illness in 0.5	Store in ventilated place away from fuming nitric acid and oxidizing materials	Use gas masks in entering atmospheres containing hydrogen sulfide. If, how-	Encountered in chemical laboratories, metallurgical and smelting works, gas

Name	Usual Shipping Container	Fire Hazard	Life Hazard	Storage	Fire-Fighting Phases	Remarks
		gen. Explosive range in air (upward propagation) 4.3 (low limit) to 46 (J). Heavier than air. Specific gravity 1.19 (air = 1). Ignition temperature 346–379°C. (655–714° F.) (6)	to 1 hour (6). Should be used under hoods in chemical laboratories to avoid breathing dangerous concentrations		ever, concentration is high, use oxygen helmet of type approved for such use by U.S. Bureau of Mines	works, sewers
Lead nitrate	Wooden barrels	Oxidizing material. Classes with sodium nitrate	Poisonous	Isolate; safeguard against mechanical injury		
Lead sulfocyanate	Fiber and stainless steel drums	Slow-burning. Products of decomposition by heat in presence of air include sulfur, carbon disulfide, and nitrogen	Poisonous when taken internally. Products of combustion contain sulfur dioxide	Store in dry place away from oxidizing materials	Water is a good extinguishing agent	
Lime (unslaked)	See calcium oxide					
Magnesium	Shavings or powder in tightly closed metal or metal-lined containers. Ingots and bars in ordinary boxes	Combustible, particularly in form of powder, shavings, or thin sheets. When powder is disseminated in air, explodes by spark. In finely-divided form liberates hydrogen in contact with water. In massive form (ingots or blocks) comparatively difficult to ignite	Serious under fire conditions. Danger of explosion and from flying particles. Do not attempt to smother unless at a safe distance, or protection is provided for eyes and face	Store remote from water or moisture, oxidizing materials, chlorine, bromine, iodine, acids, and alkalies	Smother with an excess of dry graphite. Dry sand may be used on small fires. Not advisable to use sand around machinery. Do not use water, foam, carbon tetrachloride, or carbon dioxide	
Magnesium al-	In compact or bulk		Protect eyes and	Store dust, shav-	Smother with ex-	Detailed safety

Substance	Container	Hazard	Effect	Storage	Fire	Precautions
loys (high percentage of magnesium)	for (castings, plates, etc.) difficult to ignite. Readily combustible in form of dust, turnings, and hazardous in such form with chlorine, iodine, bromine, oxidizing agents, acids, and alkalies		skin from flying particles in case of fire	ings, and turnings in metal containers in detached building or fire&resistive room	cess of dry graphite. Dry sand may be used on small fires. Not advisable to use sand around machinery. Do not use water, foam, carbon tetrachloride or carbon dioxide	precautions for handling are usually supplied by manufacturers of magnesium alloys
Magnesium nitrate	Wooden boxes	Oxidizing material. Classes with sodium nitrate		See sodium nitrate	See sodium nitrate	See sodium nitrate
Muriatic acid	See hydrochloric acid					
Naphthalene	Tins, barrels, and burlap bags	Gives off flammable vapors when heated. Flash point 176°F. Naphthalene dust formes explosive mixtures with air. Ignition temperature 559°C. (1038°F.)	Irritant. Slight narcotic effect on skin. The hot vapors produce itching, pain, and eczema (5)	Isolate; keep away from fire or heat	Water is best extinguishing agent	Foam or water applied to molten naphthalene at temperatures over 230°F. will cause foaming
Nickel nitrate	Wooden kegs	Oxidizing material. Classes with sodium nitrate	Poisonous when taken internally	Safeguard against mechanical injury		
Nitric acid	Carboys and glass bottles	May cause ignition when in contact with combustible materials; corrodes iron or steel; may cause explosion when in contact with hydrogen sulfide and certain other chemicals	Corrosive; causes severe burns by contact; deadly if inhaled	Safeguard against mechanical injury of containers; isolate from turpentine, combustible materials, carbides, metallic powders, fulminates, picrates, or chlorates		
Nitraniline or	Wooden kegs	In presence of	Poisonous	Store in dry place;		

Name	Usual Shipping Container	Fire Hazard	Life Hazard	Storage	Fire-Fighting Phases	Remarks
nitro aniline		moisture causes nitration of organic materials and may result in spontaneous ignition		safeguard against mechanical injury		
Nitrochlorobenzene	Wooden kegs	Gives off flammable vapors when heated which may form explosive mixtures with air	Serious under fire conditions	Isolate, preferably in open; if inside should be in unheated compartment or building		
Phenol		When heated yields flammable vapors. Flash point 172.4°F.	Poisonous	Soluble in water. Never store with or above food products		
Phosphorus, red	Hermetically sealed tin cans inside of wooden boxes	Flammable. Explosive when mixed with oxidizing materials	Yields toxic fumes when burning	Isolate from other chemicals; safeguard against mechanical injury of container	Flood with water, and when fire is extinguished cover with wet sand or dirt. Under certain conditions at high temperatures reverts to white phosphorus	Not as dangerous to handle as white phosphorus, and when afire, more readily extinguished
Phosphorus, white (or yellow)	Under water usually in hermetically sealed cans enclosed in other hermetically sealed cans with outside wooden boxes, or in drums or tank cars	Highly flammable. Explosive in contact with oxidizing material. Ignites spontaneously on contact with air	Poisonous. Serious under fire conditions. Yields highly toxic fumes when burning. Contact of phosphorus with skin causes severe burns	Isolate from chemicals. Store large quantities under water in underground iron or concrete tanks	Deluge with water until fire is extinguished; then cover with wet sand or dirt	
Phosphorus pentasulfide	Glass bottles and sealed drums	Slow-burning. Readily ignited by small flames. Ignition temperature about 287° C. (548.6°F.). Danger of spontaneous heating	Reacts with water, evolving hydrogen sulfide. Products of combustion include sulfur dioxide and phosphorus pentox-	Store in sealed containers away from oxidizing materials	Carbon dioxide is an effective extinguishing agent. Water is also effective but reacts with unburned material, forming hydro-	Used in synthesis of organic chemicals. Has a peculiar odor and is very hygroscopic

Material	Storage containers	Properties	Hazard to life	Safeguards	Fire fighting	Remarks
		in presence of moisture	ide. Poisonous when taken internally		gen sulfide	
Phosphorus sesquisulfide	Wooden boxes, iron drums, glass bottles	Highly flammable. Ignites by friction	Fumes in fire toxic	Isolate from chemicals. Safeguard container against shock		
Picric acid	Wooden kegs, boxes, bottles	Flammable, explosive. Oxidizing material	Classes with high explosives in respect to life hazard	Isolate or store under water, keep away from other material, including metals, with which it forms sensitive and explosive picrates		
Potassium (metallic potassium)	Hermetically sealed steel drums, tin cans, and tank cars	Oxidizes rapidly on exposure to atmosphere, igniting spontaneously if warm enough Water is decomposed suddenly by contact with potassium, sufficient heat being generated to ignite spontaneously evolved hydrogen (in presence of air). Reaction with water more violent than that of sodium	Strong caustic reaction. Dangerous	Do not tier if it can be avoided. Keep away from water, avoiding sprinkler systems. Safeguard against mechanical injury of containers	Smother with excess of dry graphite or dry sand Do not use water. See *Remarks*	It is difficult to extinguish fires in large quantities of potassium
Potassium chlorate	Wooden brarels or kegs	Oxidizing material: explosive when in contact with combustible material	Dangerous unerd fire conditions	Isolate from combustible materials, acids, and sulfur	Water is best extinguishing agent	
Potassium cyanide	Tightly closed glass, earthenware, or metal containers; wooden boxes with inside	Cyanides not flammable but evolve hydrocyanic acid (see) on contact with acids or moisture	Highly poisonous when taken internally. Evolves hydrocyanic acid gas (poisonous) on	Isolate. Safeguard containers against mechanical injury		

Name	Usual Shipping Container	Fire Hazard	Life Hazard	Storage	Fire-Fighting Phases	Remarks
Potassium hydroxide	metal containers, or with hermetically sealed metal lining; metal barrels or drums	Generates heat on contact with water. Classes with calcium oxide (lime) in hazard	contact with acids or moisture	Store in dry place; keep remote from water or moisture		See sodium nitrate
Potassium nitrate	Wooden barrels, glass bottles; Bags, tins, and glass bottles	In contact with organic materials causes violent combustion on ignition. Classes with sodium nitrate		Store in dry place, prevent contact with organic material	See sodium nitrate	
Potassium perchlorate	Paper-lined metal containers	Oxidizing material. Combustible in contact with organic materials. More stable than chlorates. Explosive in contact with concentrated sulfuric acid		Store in dry place away from acids and combustible material	Use water to prevent spread of fire	
Potassium permanganate	Tins	Oxidizing material. Explosive when treated with sulfuric acid, and in contact with alcohol, ether, flammable gases, and combustible materials		Isolate from other chemicals, especially those noted under fire hazard		
Potassium peroxide	Tins and steel drums	Does not burn or explode per se but mixtures of potassium peroxide and com-	Strong caustic reaction and dangerous under fire conditions. Avoid breathing	Store remote from organic substances and water. Do not expose to sprinkler	Smother with dry sand, soda ash, or rock dust. Do not use water	

		...bustible substances are explosive and ignite easily even by friction or on contact with a small amount of water. Reacts vigorously with water and in large quantities reaction may be explosive	dust in handling and wear goggles to protect eyes	systems	
Potassium persulfate	Glass bottles and stone jars	Oxidizing material. May cause explosion in fire		Keep dry; safeguard against mechanical injury of containers	
Potassium sulfide	Iron drums, cans, glass bottles	Moderately flammable, yields hydrogen sulfide on contact with mineral acids and sulfur dioxide when burning	Yields irritating and corrosive gases when burning	Safeguard against mechanical injury of containers	
Salicylic acid	Bottles, cartons, kegs, and barrels	Combustible solid. Flash point 157° C. (315°F.). Ignition temperature 545°C. (1013°F.). Salicylic dust forms explosive mixtures with air		Store in dry place	Slightly soluble in water. Extinguish with water or smother with carbon dioxide or sand. Used in manufacture of aspirin, salol, and methyl salicylate; also in manufacture of azo dyes. Preservative. There have been explosions in sublimation chambers
Saltpeter (see potassium nitrate)					
Silver nitrate	Amber or black glass bottles	Oxidizing material	Corrosive and poisonous	Store in dark place; keep cool and away from combustible material	

Name	Usual Shipping Container	Fire Hazard (primary class)	Life Hazard	Storage	Fire-Fighting Phases	Remarks
Silver picrate		High explosive (primary class)	Poisonous when taken internally	Explosives restrictions		Used in diluted form as antiseptic and prophylactic
Soda caustic (see sodium hydroxide)						
Sodium	Hermetically sealed steel drums, tin cans, and tank cars	Water is suddenly decomposed by contact with sodium with the evolution of hydrogen, which may ignite spontaneously (in presence of air). Classes with potassium in respect to fire hazard but its reaction with water is less violent than that of potassium	Strong caustic reaction. Dangerous	Do not tier if it can be avoided. Keep away from water, avoiding sprinkler systems. Safeguard against mechanical injury of containers	Smother with an excess of dry graphite or dry sand Do not use water. See Remarks	It is difficult to extinguish fires in large quantities of sodium
Sodium chlorate	Wooden barrels, glass bottles	Oxidizing material. Classes with potassium chlorate. See potassium chlorate	See potassium chlorate	See potassium chlorate	See potassium chlorate	
Sodium chlorite (NaClO$_2$)	Wooden boxes with inside containers which must be glass or earthenware not over 2.25-lb. capacity, or metal not over 5-lb. capacity each	Strong oxidizing material. Decomposes with heat evolved of at about 175°C. (347°F.). Explosive in contact with combustible material. See potassium chlorate. In contact with strong acid liberates chlorine dioxide, an extra-hazardous gas	Poisonous when taken internally. Dangerous under fire conditions	Isolate from combustible material, sulfur, and acids	Soluble in water. Water is best extinguishing agent	Used in bleaching textiles and paper

Material	Nature/Hazard	Containers	Storage	Fire fighting	Remarks
Sodium cyanide (see potassium cyanide)					
Sodium hydrosulfite	Combustible. Heats spontaneously in contact with moisture and air, and may ignite nearby combustible material	Wooden barrels, kegs, or boxes with inside glass bottles of capacity not exceeding 5 lb. each, or metal containers	Store in dry place away from combustible materials	Smother with sand or foam	Bleaching agent for removing dyes
Sodium hydroxide	Classes with potassium hydroxide and calcium oxide	Iron drums	Isolate from heat and water. See calcium oxide hydroxide		
Sodium nitrate	Oxidizing material. Bags or barrels may become impregnated with nitrate, in which condition they are readily ignitable. In contact with organic or other readily oxidizable (combustible) substances it will cause violent combustion on ignition	Bags, tins, and glass bottles	Store in dry place; prevent contact with organic or combustible material	Most fires involving sodium nitrate can safely be fought with water in early stages; at such times should be flooded with water. When large quantities are involved in fire, sodium nitrate may fuse or melt, in which condition application of water may result in extensive scattering of molten material; therefore, care should be taken in applying water to material after fire has been burning for some time	Fire hazard less if removed from bags and stored in noncombustible bins
Sodium perchlorate	See potassium perchlorate	Paper-lined metal chlorate	See potassium perchlorate	See potassium perchlorate	Sodium perchlorate in anhydrous form is very hygroscopic and not used

Name	Usual Shipping Container	Fire Hazard	Life Hazard	Storage	Fire-Fighting Phases	Remarks
Sodium peroxide (see potassium peroxide)						much in industry
Sodium sulfide	Iron drums and bottles	Moderately flammable. Classes with potassium sulfide	See potassium sulfied	See potassium sulfide		
Strontium nitrate	Barrels and boxes	Oxidizing material. Classes with sodium nitrate		Safeguard against mechanical injury; keep away from other materials	See sodium nitrate	
Strontium oxalate	Metal cans	Slow-burning	Poisonous. Decomposes by heat, evolving carbon monoxide and carbon dioxide	Store in dry place away from oxidizing materials	Water is good extinguishing agent	Used in manufacture of pyrotechnics
Strontium peroxid (SrO₂)	Metal cans	Oxidizing material. Hazards in class with barium peroxide		Store in dry place away from combustible materials	Smother with sand, ashes, or rock dust. Do not use water	
Sulfur	Sacks, boxes, barrels, and box cars	Flammable. Dust or vapor forms explosive mixtures with air. Hazardous in contact with oxidizing material	When burning forms sulfuric dioxide, which in concentrations of 0.2% by volume in air may cause serious injury in 0.5 hour or less (12)	Provide good ventilation. Isolate from chlorates, nitrates, and other oxidizing materials	Water in form of spray best extinguisher. Small fires may be smothered with sand or additional sulfur. (Sulfur dioxide does not support combustion.) Avoid use of pressure hose streams, and do not scatter sulfur dust	If spray nozzle is not available, water may be allowed to flow out of hose (without nozzle) on to a burning pool of sulfur, or saturated steam may be used. See N.F.P.A. Code for Prevention of Sulphur Dust Explosions and Fires
Sulfuric acid	Carboys, iron drums, glass bottles, and tank cars	May cause ignition by contact with combustible material. Corrodes metal	Corrosive; dangerous fumes under fire conditions	Safeguard against mechanical injury, isolate from saltpeter, metallic powders, carbides,	Smother with sand, ashes, or rock dust, but avoid water	

Note: SrO_2

Material	Containers	Properties / Hazards		Storage	Fire / Extinguishing	Remarks
Thorium nitrate	Wooden kegs	Oxidizing material. Classes with sodium nitrate		picrates, fulminates, chlorates, and combustible materials. Store in dry place, remote from water or moisture	See sodium nitrate	
Uranium nitrate	Glass bottles, boxes	Oxidizing material. Classes with sodium nitrate		See sodium nitrate	See sodium nitrate	
Vinyl ether	Glass bottle and metal cans	Highly volatile flammable liquid. Flash point below −22°F. Gives off even at comparatively low temperatures vapors which form flammable mixtures with air or oxygen. Explosive range 1.70 to 27.0% (upward propagation (4). Hazard in class with ethyl ether	Anesthetic	Isolate. Safeguard containers against mechanical injury. Store in a cool, well-ventilated storeroom	Lighter than water (sp. gr. 0.774). Not soluble in water. Water may be used only to cool metal containers. Best extinguishing agents are carbon dioxide, sand, and carbon tetrachloride	Only electrical equipment of explosion-proof type. Group C classification, permitted in atmospheres containing vinyl ether vapor in flammable proportions. Group C classification for vinyl ether tentative pending further tests
Zinc chlorate	Glass bottles, iron drums	When in contact with organic material explodes by slight friction, percussion or shock. Classes with potassium chlorate	Serious under fire conditions	Safeguard against mechanical injury; avoid tiering; isolate		
Zinc powder or dust	Cartons, wooden barrels, or steel drums	Hydrogen evolved when commercial zinc is in contact with acids, sodium hydroxide, or potassium hydroxide. Hydrogen also evolved by acid-forming	Zinc comparatively volatile at elevated temperatures. Under fire conditions precautions should be taken to avoid breathing fumes, which may cause	Store in dry, ventilated place away from water or moisture. Isolate from acids	Smother with sand, ashes, or rock dust. Do not use water	

Name	Usual Shipping Container	Fire Hazard	Life Hazard	Storage	Fire-Fighting Phases	Remarks
		combinations containing zinc, such as zinc chloride and moisture. Dust may form explosive mixtures with air. Zinc dust in bulk in damp state may heat and ignite spontaneously on exposure to air	metal fume fever			
Zirconium	Wooden kegs, glass bottles	Has comparatively low ignition temperature. Highly flammable in dry state. Burns with intensely brilliant flame. Explosive in contact with oxidizing agents. Powder very susceptible to ignition by static electricity, and explosion may be caused when dispersed into a cloud in air by static charges generated		Store only in wet condition and in small quantities. Isolate from oxidizing materials	Investigation of this phase not complete, but available data indicate fires can be controlled by foam or sand. Carbon tetrachloride, carbon dioxide, soda and acid extinguishers ineffective	Encountered in granular, finely divided powder; also in form of small, friable, spongy lumps

1. Bur. Mines, *Bull.,* 279 (1931).
2. Bur. Mines, *Rept. Investigations,* 3278 (1935).
3. Bur. Mines, *Tech. Paper,* 544 (1933).
4. *Chem. Reviews,* 55, 4 (1938).
5. Flury, F., and Zernik, F., "Schadliches Gase," Berlin, Julius Springer, 1931.
6. International Critical Tables, New York, McGraw-Hill Book Co.
7. International Labour Office, Geneva, "Occupation and Health," 1930.
8. *J. Chem. Soc.,* 154, 1688 (1922).
9. *J. Ind. Hyg.,* 18, 459 (1936).
10. Underwriters Laboratories, Inc.
11. Underwriters Laboratories, "Methods for Classification of Hazards of Liquids."
12. Underwriters Laboratories, "Report on Comparative Life and Explosion Hazards of Common Refrigerants."
13. Underwriters Laboratories, *Report,* MH 1646 (1927).
14. Underwriters Laboratories, "Report on Fire Hazard of Ethane, Propane, Butane, and Ammonia as Refrigerants."

TABLES

Conversion Factors

1. Grams per liter (g./l.) multiplied by 0.134 = avoirdupois ounces per gallon (oz./gal.).

2. Avoirdupois ounces per gallon (oz./gal.) multiplied by 7.5 = grams per liter (g./l.).

3. Grams per liter (g./l.) multiplied by 0.122 = troy ounces per gallon (troy oz./gal).

4. Troy ounces per gallon (troy oz./gal.) multiplied by 8.2 = grams per liter (g./l.).

5. Grams per liter (g./l.) multiplied by 2.44 = pennyweights per gallon (dwt./gal.).

6. Pennyweights per gallon (dwt./gal.) multiplied by 0.41 = grams per liter (g./l.).

7. Amperes per square decimeter (amp./dm.^{2}) multiplied by 9.29 = amperes per square foot (amp./sq. ft.).

8. Amperes per square foot (amp./sq. ft.) multiplied by 0.108 = amperes per square decimeter (amp./dm.^{2}).

Thermometer Readings:

Degrees Centigrade $\times$ 1.8 + 32 = degrees Fahrenheit

$$\frac{\text{Degrees Fahrenheit} - 32}{1.8} = \text{degrees Centigrade}$$

$$\frac{\text{Degrees Reaumur} \times 9}{4} + 32 = \text{degrees Fahrenheit}$$

$$\frac{(\text{Degrees Fahrenheit} - 32)4}{9} = \text{degrees Reaumur}$$

$$\frac{\text{Degrees Reaumur} \times 5}{4} = \text{degrees Centigrade}$$

$$\frac{\text{Degrees Centigrade} \times 4}{5} = \text{degrees Reaumur}$$

Specific Gravity
Weight Required to Make a Gallon

	Specific Gravity	Pounds to Gallon
Litharge	9.3	77.5
Red Lead	8.7 to 8.8	72.5
Orange Mineral (orange lead)	8.6 to 8.7	73.0
White-Lead	6.7	55.8
Basic Lead Sulfate	6.4	53.3
Chrome Yellow (medium)	6.0	50.0
Zinc Oxide (white zinc)	5.6	46.6
Basic Lead Chromate	6.8	56.6
English (mercury) Vermilion	8.2	68.3
Bright Red Oxide of Iron	4.9 to 5.26	42.0
Indian Red Oxide of Iron	5.26	43.8
Brown Oxide of Iron (Prince's)	3.2	26.6
Ultramarine	2.4	20.0
Prussian Blue	1.85	15.4
Chrome Green (blue tone)	4.44	37.0

	Specific Gravity	Pounds to Gallon
Chrome Green (yellow tone)	4.0	33.0
Lithopone	4.25	35.4
Ochre	2.94	24.5
Barytes ....4.35 to	4.46	35 to 37.0
Blanc Fixe	4.25	35.4
Gypsum (terra alba)	2.3	19.0
Asbestine (magnesium silicate)	2.75	23.0
China Clay (aluminum silicate) .... 2.6 to	2.7	22.5
Whiting	2.65	22.0
Silica	2.65	22.0
Natural Graphite .... 2.1 to	2.4	18.0
Acheson's Graphite	2.2	18.3
Lampblack	1.85	15.4
Carbon Black	1.85	15.4
Keystone Filler (ground slate)	2.66	22.0
Titanox	4.3	35.8
Titanium Oxide .... 3.9 to	4.0	33.3
Drop Black	2.5	20.8

To this table the following data may be added: The weight of one gallon of paste made with

	Pounds
Red Lead..	44.8
White Lead (heavy paste)	34.0
White Lead (soft paste).......................................	30.8
White Zinc	25.0
Chrome Yellow (medium)	24.0
Chrome Green	24.0
Venetian Red	19.0
French Ochre	15.0
Prussian Blue	10.0
Lampblack	9.1
Drop Black	11.7

Weights and Measures
English System
Avoirdupois and Commercial Weights

16 drams, or 437.5 grains	= 1 ounce, oz.
16 ounces, or 7000 grains	= 1 pound, lb.
28 pounds	= 1 quarter, qr.
4 quarters (English)	= 1 hundredweight, cwt., 112 lbs.
20 hundredweight	= 1 ton of 2240 lbs., gross or long ton
2000 pounds	= 1 net, or short, ton
2204.6 pounds	= 1 metric ton = 1000 kilos

1 stone = 14 pounds; 1 quintal = 100 pounds

Troy Weights

24 grains	= 1 pennyweight, dwt.
20 pennyweights	= 1 ounce, oz. = 480 grains
12 ounces	= 1 pound, lb. = 5760 grains
1 carat	= 3.168 grains = 0.205 gram

Troy weight is used for weighing gold and silver. The grain is the same in Avoirdupois, Troy and Apothecaries' weights.

Apothecaries' Weights

20 grains	= 1 scruple
2 scruples	= 1 drachm, 3 = 60 grains
8 drachms	= 1 ounce, 3 = 480 grains
12 ounces	= 1 pound, lb. = 5760 grains

Apothecaries' Measures

60 minims (min.)	= 1 fluid drachm (fl. dr.)
8 fluid drachms	= 1 fluid ounce (fl. oz.)
16 fluid ounces	= 1 pint (O)
8 pints	= 1 gallon (C)

Relations of Apothecaries' Measures to Weights
(All liquids to be measured at 62°F.)

1 inim is the m measure of	0.0115	grains of distilled water
1 fluid drachm " "	54.687	" " " "
1 fluid ounce " "	437.5	" " " "
1 pint " "	8750	" " " "
1 gallon " "	70000	" " " "

Linear Measure

12 inches	= 1 foot	4 poles	= 1 chain
3 feet	= 1 yard	40 poles	= 1 furlong
6 feet	= 1 fathom	8 furlongs	= 1 mile = 1760 yards
5½ yards	= 1 rod pole, or perch		

Square Measure

144 square inches = 1 square foot
9 square feet = 1 square yard
30.25 square yards or 272.5 sq. feet = 1 square rod
160 square rods or 4840 sq. yards or 43560 sq. feet = 1 acre
640 acres = 1 square mile
An acre = a square whose side is 208.7 feet

Cubic Measure

1728 cubic inches	= 1 cubic foot
27 cubic feet	= 1 cubic yard

1 cord of wood = a pile 4 × 4 × 8 feet = 128 cubic feet
1 perch of masonry = 16.5 × 1.5 × 1 foot = 24.75 cubic feet
1 cubic inch of water at 62°F. weighs 252.286 grains

" " " " " " "	0.57665 oz. (av.)
" " " " " " "	0.036041 lb.
1 cubic foot " " " " "	996.458 oz. (av.)
" " " " " " "	62.2786 lb.
1 cubic yard " " " "	0.75068 tons

Capacity Measure

Liquid

4 gills = 1 pint
2 pints = 1 quart
4 quarts = 1 gallon

Relation of Capacity, Volume and Weight

1 pint   = 28.875 cubic inches
1 quart   = 57.75 cubic inches
1 gallon (U.S.)   = 231 cubic inches
1 gallon (English)   = 277.274 cubic inches
7.4805 gallons   = 1 cubic foot
1 gallon water at 62 F. weighs 8.3356 lbs.

Dry

2 pints   = 1 quart
8 quarts   = 1 peck
4 pecks   = 1 bushel
1 U. S. standard bushel (struck)   = 2150.42 cubic inches
0.80356 U. S. bushels (struck)   = 1 cubic foot

Metric Equivalents

Linear Measure

1 centimeter = 0.3937 in.
1 decimeter = 3.937 in. = 0.328 ft.
1 meter = 39.37 in. = 1.0936 yds.
1 decameter = 1.9884 rods
1 kilometer = 0.62137 miles
1 inch = 2.54 centimeters
1 foot = 3.048 decimeters
1 yard = 0.9144 meters
1 rod = 0.5029 decameters
1 mile = 1.6093 kilometers
The meter, as used in Europe, is 39.370432 inches.
1 sq. foot = 9.2903 sq. decimeters
1 sq. yard = 0.8361 sq. meters
1 sq. rod = 0.2529 ares
1 acre = 0.4047 hectares
1 sq. mile = 0.259 sq. kilometers

Square Measure

1 sq. centimeter = 0.1550 sq. inches
1 sq. decimeter = 0.1076 sq. feet
1 sq. meter = 1.196 sq. yards.
1 are = 3.954 sq. rods
1 hectare = 2.47 acres
1 sq. kilometer = 0.386 sq. miles
1 sq. inch = 6.452 sq. centimeters

Approximate Metric Equivalents

1 decimeter = 4 inches
1 meter = 1.1 yards
1 kilometer = 5/8 of a mile
1 hectare = 2½ acres
1 stere, or cu. meter = ¼ of a cord
1 liter = 1.06 qt. liquid, 0.9 qt. dry
1 kilogram = 2⅕ lbs.
1 metric ton = 2200 lbs.

Weights

1 decigram = 0.003527 oz. = 1.5432 grains
1 gram = 0.03527 oz. avoir., or about 15½ troy grains
1 kilogram = 2.2046 lbs. avoir.
1 metric ton = 1.1023 English short tons
1 ounce avoir. = 28.35 grams
1 pound avoir. = 0.4536 kilograms
1 English short ton = 0.9072 metric tons

Comparison of Avoirdupois and Metric Weights

Grains	Drams Av.	Oz. Av.	Lbs. Av.	Deniers	Grams
1.000	......	......	......	1.296	0.065
27.340	1.000	......	......	35.437	1.772
437.500	16.000	1.000	......	566.990	28.350
7000.000	256.000	16.000	1.000	9071.840	453.840
0.772	......	......	......	1.000	0.050
15.432	......	0.03527	......	20.000	1.000

Solution Strength	pH Values of Chemicals Reagent	pH
1% Commercial Olive Oil Soap (Neutral)		10.1 — 10.3
1% Commercial Olive Oil Soap (Neutral)		10.1 — 10.3
1% Commercial Olive Oil or Tallow Soap Containing 20% Soda Ash		10.75 — 10.88
1% Commercial Olive Oil or Tallow Soap Containing 5% Caustic		12.0 — 12.2
½% Commercial Olive Oil or Tallow Soap		10.0 — 10.2
¼% Commercial Olive Oil or Tallow Soap		9.9 — 10.1
1% Sulfonated Oils (Neutral)		6.0 — 7.0
1% Sulfonated Oils Containing Free Acid		Below 6.0
1% Sulfonated Oils Containing Soap or Alkalies		Above 7.0
¼% Trisodium Phosphate		12.3
¼% Sodium Silicate		12.2
¼% Sodium Carbonate		11.3
¼% Sodium Sulphite		9.7
¼% Disodium Phosphate		8.9
¼% Borax		8.8
¼% Monosodium Phosphate		5.0

THE HANDBOOK OF CHEMICAL SUBSTITUTES

Equivalents of Twaddell, Baumè and Specific Gravity Scales

Twaddell	Baumé	Specific Gravity	Twaddell	Baumé	Specific Gravity	Twaddell	Baumé	Specific Gravity	Twaddell	Baumé	Specific Gravity
0	0	1.000	44	26.0	1.220	88	44.1	1.440	131	57.1	1.655
1	0.7	1.005	45	26.4	1.225	89	44.4	1.445	132	57.4	1.660
2	1.4	1.010	46	26.9	1.230	90	44.8	1.450	133	57.7	1.665
3	2.1	1.015	47	27.4	1.235	91	45.1	1.455	134	57.9	1.670
4	2.7	1.020	48	27.9	1.240	92	45.4	1.460	135	58.2	1.675
5	3.4	1.025	49	28.4	1,245	93	45.8	1.465	136	58.4	1.680
6	4.1	1.030	50	28.8	1.250	94	46.1	1.470	137	58.7	1.685
7	4.7	1.035	51	29.3	1.255	95	46.4	1.475	138	58.9	1.690
8	5.4	1.040	52	29.7	1.260	96	46.8	1.480	139	59.2	1.695
9	6.0	1.045	53	30.2	1.265	97	47.1	1.485	140	59.5	1.700
10	6.7	1.050	54	30.6	1.270	98	47.4	1.490	141	59.7	1.705
11	7.4	1.055	55	31.1	1.275	99	47.8	1.495	142	60.0	1.710
12	8.0	1.060	56	31.5	1.280	100	48.1	1.500	143	60.2	1.715
13	8.7	1.065	57	32.0	1.285	101	48.4	1.505	144	60.4	1.720
14	9.4	1,070	58	32.4	1.290	102	48.7	1.510	145	60.6	1.725
15	10.0	1.075	59	32.8	1.295	103	49.0	1.515	146	60.9	1.730
16	10.6	1.080	60	33.3	1.300	104	49.4	1.520	147	61.1	1.735
17	11.2	1.085	61	33.7	1.305	105	49.7	1.525	148	61.4	1.740
18	11.9	1.090	62	34.2	1.310	106	50.0	1.530	149	61.6	1.745
19	12.4	1.095	63	34.6	1.315	107	50.3	1.535	150	61.8	1.750
20	13.0	1.100	64	35.0	1.320	108	50.6	1.540	151	62.1	1.755
21	13.6	1.105	65	35.4	1.325	109	50.9	1.545	152	62.3	1.760
22	14.2	1.110	66	35.8	1.330	110	51.2	1.550	153	62.5	1.765
23	14.9	1.115	67	36.2	1.335	111	51.5	1.555	154	62.8	1.770
24	15.4	1.120	68	36.6	1.340	112	51.8	1.560	155	63.0	1.775
25	16.0	1.125	69	37.0	1.345	113	52.1	1.565	156	63.2	1.780
26	16.5	1.130	70	37.4	1.350	114	52.4	1.570	157	63.5	1.785
27	17.1	1.135	71	37.8	1.355	115	52.7	1.575	158	63.7	1.790
28	17.7	1.140	72	38.2	1.360	116	53.0	1.580	159	64.0	1.795
29	18.3	1.145	73	38.6	1.365	117	53.3	1.585	160	64.2	1.800
30	18.8	1.150	74	39.0	1.370	118	53.6	1.590	161	64.4	1.805
31	19.3	1.155	75	39.4	1.375	119	53.9	1.595	162	64.6	1.810
32	19.8	1.160	76	39.8	1.380	120	54.1	1.600	163	64.8	1.815
33	20.3	1.165	77	40.1	1.385	121	54.4	1.605	164	65.0	1.820
34	20.9	1.170	78	40.5	1.390	122	54.7	1.610	165	65.2	1.825
35	21.4	1.175	79	40.8	1.395	123	55.0	1.615	166	65.5	1.830
36	22.0	1.180	80	41.2	1.400	124	55.2	1.620	167	65.7	1.835
37	22.5	1.185	81	41.6	1.405	125	55.5	1.625	168	65.9	1.840
38	23.0	1.190	82	42.0	1.410	126	55.8	1.630	169	66.1	1.845
39	23.5	1.195	83	42.3	1.415	127	56.0	1.635	170	66.3	1.850
40	24.0	1.200	84	42.7	1.420	128	56.3	1.640	171	66.5	1.855
41	24.5	1.205	85	43.1	1.425	129	56.6	1.645	172	66.7	1.860
42	25.0	1.210	86	43.4	1.430	130	56.9	1.650	173	67.0	1.865
43	25.5	1.215	87	43.8	1.435						

Atomic Weights

Actinium	Ac	89	(227)	Iridium	Ir	77	192.2
Aluminum	Al	13	26.9815	Iron	Fe	26	55.847
Americium	Am	95	(243)	Krypton	Kr	36	83.80
Antimony	Sb	51	121.75	Lanthanum	La	57	138.91
Argon	Ar	18	39.948	Lawrencium	Lw	013	(256)
Arsenic	As	33	74.9216	Lead	Pb	82	207.19
Astatine	At	85	(210)	Lithium	Li	3	6.939
Barium	Ba	56	137.34	Lutetium	Lu	71	174.97
Berkelium	Bk	97	(247)	Magnesium	Mg	12	24.312
Beryllium	Be	4	9.0122	Manganese	Mn	25	54.9380
Bismuth	Bi	83	208.980	Mendelevium	Md	101	(256)
Boron	B	5	10.811	Mercury	Hg	80	200.59
Bromine	Br	35	79.909	Molybdenum	Mo	42	95.94
Cadmium	Cd	48	112.40	Neodymium	Nd	60	144.24
Calcium	Ca	20	40.08	Neon	Ne	10	20.183
Californium	Cf	98	(251)	Neptunium	Np	93	(237)
Carbon	C	6	12.01115	Nickel	Ni	28	58.71
Cerium	Ce	58	140.12	Niobium	Nb	41	92.906
Cesium	Cs	55	132.905	Nitrogen	N	7	14.0067
Chlorine	Cl	17	35.453	Osmium	Os	76	190.2
Chromium	Cr	24	51.996	Oxygen	O	8	15.9994
Cobalt	Co	27	58.9332	Palladium	Pd	46	106.4
Copper	Cu	29	63.54	Phosphorus	P	15	30.9738
Curium	Cm	96	(247)	Platinum	Pt	78	195.09
Dysprosium	Dy	66	162.50	Plutonium	Pu	94	(244)
Einsteinium	Es	99	(254)	Polonium	Po	84	(209)
Element 102		102	(254)	Potassium	K	19	39.102
Erbium	Er	68	167.26	Praseodymium	Pr	59	140.907
Europium	Eu	63	151.96	Promethium	Pm	61	(145)
Fermium	Fm	100	(253)	Protactinium	Pa	91	(231)
Fluorine	F	9	18.9984	Radium	Ra	88	(226)
Francium	Fr	87	(223)	Radon	Rn	86	(222)
				Rhenium	Re	75	186.2
Gadolinium	Gd	64	157.25	Rhodium	Rh	45	102.905
Gallium	Ga	31	69.72	Rubidium	Rb	37	85.47
Germanium	Ge	32	72.59	Ruthenium	Ru	44	101.07
Gold	Au	79	196.967	Samarium	Sm	62	150.35
Hafnium	Hf	72	178.49	Scandium	Sc	21	44.956
Helium	He	2	4.0026	Selenium	Se	34	78.96
Holmium	Ho	67	164.930	Silicon	Si	14	28.086
Hydrogen	H	1	1.00797	Silver	Ag	47	107.870
Indium	In	49	114.82	Sodium	Na	11	22.9898
Iodine	I	53	126.9044	Strontium	Sr	38	87.62

Atomic Weights (cont'd.)

Sulfur	S	16	32.064	Titanium	Ti	22	47.90
Tantalum	Ta	73	180.948	Tungsten	W	74	183.85
Technetium	Tc	43	(97)	Uranium	U	92	238.03
Tellurium	Te	52	127.60	Vanadium	V	23	50.942
Terbium	Tb	65	158.924	Xenon	Xe	54	131.30
Thallium	Tl	81	204.37	Ytterbium	Yb	70	173.04
Thorium	Th	90	232.038	Yttrium	Y	39	88.905
Thulium	Tm	69	168.934	Zinc	Zn	30	65.37
Tin	Sn	50	118.69	Zirconium	Zr	40	91.22

A value given in parentheses denotes the mass number of the longest-lived isotope.

Alcohol Proof and Percentage Table

U.S. Proof at 60°F.	% Alcohol by Volume at 60°F.	% Alcohol by Weight	U.S. Proof at 60°F.	% Alcohol by Volume at 60°F.	% Alcohol by Weight
0	0.0	0.00	26	13.0	10.50
1	0.5	—	27	13.5	—
2	1.0	0.80	28	14.0	11.32
3	1.5	—	29	14.5	—
4	2.0	1.59	30	15.0	12.14
5	2.5	—	31	15.5	—
6	3.0	2.39	32	16.0	12.96
7	3.5	—	33	16.5	—
8	4.0	3.19	34	17.0	13.79
9	4.5	—	35	17.5	—
10	5.0	4.00	36	18.0	14.61
11	5.5	—	37	18.5	—
12	6.0	4.80	38	19.0	15.44
13	6.5	—	39	19.5	—
14	7.0	5.61	40	20.0	16.27
15	7.5	—	41	20.5	—
16	8.0	6.42	42	21.0	17.10
17	8.5	—	43	21.5	—
18	9.0	7.23	44	22.0	17.93
19	9.5	—	45	22.5	—
20	10.0	8.05	46	23.0	18.77
21	10.5	—	47	23.5	—
22	11.0	8.86	48	24.0	19.60
23	11.5	—	49	24.5	—
24	12.0	9.68	50	25.0	20.44
25	12.5	—	51	25.5	—

U.S. Proof at 60°F	% Alcohol by Volume at 60°F.	% Alcohol by weight	U.S. Proof at 60°F.	% Alcohol by Volume at 60°F.	% Alcohol by Weight
52	26.0	21.82	100	50.0	42.49
53	26.5	—	101	50.5	—
54	27.0	22.13	102	51.0	43.43
55	27.5	—	103	51.5	—
56	28.0	22.97	104	52.0	44.37
57	28.5	—	105	52.5	—
58	29.0	23.82	106	53.0	45.33
59	29.5	—	107	53.5	—
60	30.0	24.67	108	54.0	46.28
61	30.5	—	109	54.5	—
62	31.0	25.52	110	55.0	47.24
63	31.5	—	111	55.5	—
64	32.0	26.38	112	56.0	48.21
65	32.5	—	113	56.5	—
66	33.0	27.24	114	57.0	49.19
67	33.5	—	115	57.5	—
68	34.0	28.10	116	58.0	50.17
69	34.5	—	117	58.5	—
70	35.0	28.97	118	59.0	51.15
71	35.5	—	119	59.5	—
72	36.0	29.84	120	60.0	52.15
73	36.5	—	121	60.5	—
74	37.0	30.72	122	61.0	53.15
75	37.5	—	123	61.5	—
76	38.0	31.60	124	62.0	54.15
77	38.5	—	125	62.5	—
78	39.0	32.48	126	63.0	55.16
79	39.5	—	127	63.5	—
80	40.0	33.36	128	64.0	56.18
81	40.5	—	129	64.5	—
82	41.0	34·25	130	65.0	57.21
83	41.5	—	131	65.5	—
84	42.0	35.15	132	66.0	58.24
85	42.5	—	133	66.5	—
86	43.0	36.05	134	67.0	59.28
87	43.5	—	135	67.5	—
88	44.0	36.96	136	68.0	60.32
89	44.5	—	137	68.5	—
90	45.0	37.86	138	69.0	61.38
91	45.5	—	139	69.5	—
92	46.0	38.78	140	70.0	62.44
93	46.5	—	141	70.5	—
94	47.0	39.70	142	71.0	63.51
95	47.5	—	143	71.5	—
96	48.0	40.62	144	72.0	64.59
97	48.5	—	145	72.5	—
98	49.0	41.55	146	73.0	65.67
99	49.5	—	147	73.5	—

U.S. Proof at 60°F	% Alcohol by Volume at 60°F.	% Alcohol by Weight	U.S. Proof at 60°F	% Alcohol by Volume at 60°F.	% by	Alcohol Weight
148	74.0	66.77	175	87.5		—
149	74.5	—	176	88.0		83.14
150	75.0	67.87	177	88.5		—
151	75.5	—	178	89.0		84.41
152	76.0	68.92	179	89.5		—
153	76.5	—	180	90.0		85.69
154	77.0	70.10	181	90.5		—
155	77.5	—	182	91.0		86.99
156	78.0	71.23	183	91.5		—
157	78.5	—	184	92.0		88.31
158	79.0	72.38	185	92.5		—
159	79.5	—	186	93.0		89.95
160	80.0	73.53	187	93.5		—
161	80.5	—	188	94.0		91.02
162	81.0	74.69	189	94.5		—
163	81.5	—	190	95.0		92.42
164	82.0	75.86	191	95.5		—
165	82.5	—	192	96.0		93.85
166	83.0	77.04	193	96.5		—
167	83.5	—	194	97.0		95.32
168	84.0	78.23	195	97.5		—
169	84.5	—	196	98.0		96.82
170	85.0	79.44	197	98.5		—
171	85.5	—	198	99.0		98.38
172	86.0	80.62	199	99.5		—
173	86.5	—	200	100.0		100.00
174	87.0	81.90				

Conversion of Thermometer Readings

F°	C°	F°	C°	F°	C°	F°	C°	F°	C	F°	C°
—40	—40.00	30	—1.11	80	26.67	250	121.11	500	260.00	900	482.22
—38	—38.89	31	—0.56	81	27.22	255	123.89	505	262.78	910	487.78
—36	—37.78	32	0.00	82	27.78	260	126.67	510	265.56	920	493.33
—34	—36.67	33	0.56	83	28.33	265	129.44	515	268.33	930	498.89
—32	—35.56	34	1.11	84	28.89	270	132.22	520	271.11	940	504.44
—30	—34.44	35	1.67	85	29.44	275	135.00	525	273.89	950	510.00
—28	—33.33	36	2.22	86	30.00	280	137.78	530	276.67	960	515.56
—26	—32.22	37	2.78	87	30.56	285	140.55	535	279.44	970	521.11
—24	—31.11	38	3.33	88	31.11	290	143.33	540	282.22	980	526.67
—22	—30.00	39	3.89	89	31.67	295	146.11	545	285.00	990	532.22

F"	C"	F"	C·	F"	C"	F"	C"	F	C	F·	C"
—20	—28.89	40	4.44	90	32.22	300	148.89	550	287.78	1000	537.78
—18	—27.78	41	5.00	91	32.78	305	151.67	555	290.55	1050	565.56
—16	—26.67	42	5.56	92	33.33	310	154.44	560	293.33	1100	593.33
—14	—25.56	43	6.11	93	33.89	315	157.22	565	296.11	1150	621.11
—12	—24.44	44	6.67	94	39.44	320	160.00	570	298.89	1200	648.89
—10	—23.33	45	7.22	95	35.00	325	162.78	575	301.67	1250	676.67
—8	—22.22	46	7.78	96	35.56	330	165.56	580	304.44	1300	704.44
—6	—21.11	47	8.33	97	36.11	335	168.33	585	307.22	1350	732.22
—4	—20.00	48	8.89	98	36.67	340	171.11	590	310.00	1400	760.00
—2	18.89	49	9.44	99	37.22	345	173.89	595	312.78	1450	787.78
0	—17.78	50	10.00	100	37.78	350	176.67	600	315.56	1500	815.56
1	—17.22	51	10.56	105	40.55	355	179.44	610	321.11	1550	843.33
2	—16.67	52	11.11	110	43.33	360	182.22	620	326.67	1600	871.11
3	—16.11	53	11.67	115	46.11	365	185.00	630	332.22	1650	898.89
4	15.56	54	12.22	120	48.89	370	187.78	640	337.78	1700	926.67
5	—15.00	55	12.78	125	51.67	375	190.55	650	343.33	1750	954.44
6	—14.44	56	13.33	130	54.44	380	193.33	660	348.89	1800	982.22
7	—13.89	57	13.89	135	57.22	385	196.11	670	354.44	1850	1010.00
8	—13.33	58	14.44	140	60.00	390	198.89	680	360.00	1900	1037.78
9	—12.78	59	15.00	145	62.78	395	201.67	690	365.56	1950	1065.56
10	—12.22	60	15.56	150	65.56	400	204.44	700	371.11	2000	1093.33
11	—11.67	61	16.11	155	68.33	405	207.22	710	376.67	2050	1121.11
12	—11.11	62	16.67	160	71.11	410	210.00	720	382.22	2100	1148.89
13	—10.56	63	17.22	165	73.89	415	212.78	730	387.78	2150	1176.67
14	—10.00	64	17.78	170	76.67	420	215.56	740	393.33	2200	1204.44
15	—9.44	65	18.33	175	79.44	425	218.33	750	398.89	2250	1232.22
16	—8.89	66	18.89	180	82.22	430	221.11	760	404.44	2300	1260.00
17	—8.33	67	19.44	185	85.00	435	223.89	770	410.00	2350	1287.78
18	—7.78	68	20.00	190	87.78	440	226.67	780	415.56	2400	1315.56
19	—7.22	69	20.56	195	90.53	445	229.44	790	421.11	2450	1343.33
20	—6.67	70	21.11	200	93.33	450	232.22	800	426.67	2500	1371.11
21	—6.11	71	21.67	205	96.11	455	235.00	810	432.22	2550	1398.89
22	—5.56	72	22.22	210	98.89	460	237.78	820	437.78	2600	1426.67
23	—5.00	73	22.78	215	101.67	465	240.55	830	443.33	2650	1454.44
24	—4.44	74	23.33	220	104.44	470	243.33	840	448.89	2700	1482.22
25	—3.89	75	23.89	225	107.22	475	246.11	850	454.44	2750	1510.00
26	—3.33	76	24.44	230	110.00	480	248.89	860	460.00	2800	1537.78
27	—2.78	77	25.00	235	112.78	485	251.67	870	465.56	2850	1565.56
28	—2.22	78	25.56	240	115.56	490	254.44	880	471.11	2900	1593.33
29	—1.67	79	26.11	245	118.33	495	257.22	890	476.67	2950	1621.11